Michel Büch
Whiteface

American Frictions

Editors
Carsten Junker
Julia Roth
Darieck Scott

Editorial Bord
Arjun Appadurai, New York University
Mita Banerjee, University of Mainz
Tomasz Basiuk, University of Warsaw
Isabel Caldeira, University of Coimbra

Volume 5

Michel Büch
Whiteface

Improv Comedy and Anti-Blackness

DE GRUYTER

The open access publication of this title was supported by the publication fund
of the Staats- und Universitätsbibliothek Bremen.

ISBN 978-3-11-135840-6
e-ISBN (PDF) 978-3-11-075274-8
e-ISBN (EPUB) 978-3-11-075282-3
DOI https://doi.org/10.1515/783110752748

This work is licensed under the Creative Commons Attribution-NonCommercial-NoDerivatives 4.0
International License. For details go to https://creativecommons.org/licenses/by-nc-nd/4.0/.

Library of Congress Control Number: 2021949770

Bibliographic information published by the Deutsche Nationalbibliothek
The Deutsche Nationalbibliothek lists this publication in the Deutsche Nationalbibliografie;
detailed bibliographic data are available on the Internet at http://dnb.dnb.de.

© 2023 Michel Büch, published by Walter de Gruyter GmbH, Berlin/Boston.
This volume is text- and page-identical with the hardback published in 2022.
The book is published open access at www.degruyter.com.
Printing and binding: CPI books GmbH, Leck

www.degruyter.com

Always assume I'm a white guy. Always.
In all my scenes, I am always white.
Because it's a neutral guy.
Warren Phynix Johnson

With being human, everything is praxis.
Sylvia Wynter

Acknowledgments

This project began with and is largely based on many hours of personal conversations with Chicago improvisers, who generously shared their thoughts and expertise on racism in *improv as lived* with a white scholar from Germany. Beyond the substantial contributions their vastly heterogenous perceptions of improv and racism have made to this book, many of the things I heard have impacted me personally and instigated life decisions I would otherwise have made differently. In the meantime, many of them have gone on to great careers. They include: Joel Boyd, Aasia LaShay Bullock, TJ Jagodowski, David Pasquesi, Dewayne Perkins, Warren Phynix Johnson, Derek Schleelein, Dacey Arashiba, Kimberly Michelle Vaughn, Patrick Rowland, and Loreen Targos. I spoke with others as well, but some I could not get hold of again, and some decided not to authorize my use of their material because "times have changed." (Needless to say, the group who denied authorization was homogenously white.) However, all these conversations helped me understand improv discourse better because they *are* that very discourse.

I could not have wished for a better dissertation supervisor and intellectual mentor than Sabine Broeck. Her spirited cooperation and exchange of ideas, always on eye level, her encouragement to think in all directions in terms of genre, her never-ending methodological curiosity, her unpretentious admiration of quality as much as her unflinching and immensely productive derision of pretentiousness and empty talk have made the book what it is. Further, she was faced with a doctoral candidate who was not working within the academy professionally and thus always had other things on his plate as well. Almost more than anything else, I wish to thank her for the trust and patience she had for this project. Plus, she seemed to know exactly when to remind me that this work was something rather than nothing. Without her, this book would not exist in any way, shape, or form.

I was very lucky to have not one but two interested and committed mentors. The influence that Jaye Austin Williams had on the development of this book is immense. I had the luxury of experiencing her as a teacher as well as an examiner for the dissertation. Her gift to the world is a combination of seemingly unlimited personal generosity with relentless articulations vis-à-vis the world we live in, and the art of using language rich with meaning beyond the words themselves. She can nonchalantly drop a side remark that changes one's view of talking race and walking space. Thank you.

Further, my gratitude goes out to my wider examination board – Susan Winnett, Elisabeth Arend, Norbert Schaffeld – and everyone else who attended my defense for participating in a space specifically designed for my project to resonate. I could have kept conversing for many hours more.

Throughout the process, I have benefited from various interlocuters who have read, listened, and responded to aspects and sections of this study, thus helping me develop the project. Many of them have been organized in the doctoral network *Perspektiven in der Kulturanalyse* at Bremen University. Of the many wonderful people in that network, Samira Spatzek holds a special place for me. Not only did she put great effort into pointing me in the relevant direction when it came to the theorization of property, but the two of us also had the honor of struggling through Sylvia Wynter's enormous essay "Unsettling the Coloniality of Being/Power/Truth/Freedom" as translators. Other members of the network who have engaged with my project include Mariya Nikolova, Cedric Essi, Courtney Moffett-Bateau, Marius Henderson, and Paula von Gleich, all of whom have in their very different ways helped me move on at different times throughout the project – thank you all!

There are two more people who are not only personally very dear to me but have impacted this study greatly in terms of content. I would have never considered Donald Winnicott as a theorizer of play had it not been for Kiana Ghaffarizad – an adroit academic with admirable personal integrity and a dear friend ever since I started studying at university. The second is my close companion Jörn Grebe, an outstanding psychoanalyst both as scholar and as a practitioner. Numerous conversations with him have been important for me to understand, question, and validate the psychoanalytic dimension that runs through this project.

I also wish to thank Carsten Junker and the editorial board of American Frictions for expressing first interest and then trust in this project. In addition, Julie Miess and her colleagues at De Gruyter have made me feel more than welcome. I also thank Annie Moore for her excellent editing.

I am very lucky to have been supported by both my parents, Elke Büch and Hein Göttsch-Büch, who have provided ample physical, temporal, and psychological space for me to do the actual writing. More than once their curiosity has forced me to make my points in fewer words than I used in the written elaborations.

This project also could not exist without the help and ongoing encouragement of Andrea Hingst. Her deep-seated conviction that the project was worth bringing to some sort of end, and that it should be made public, helped me through phases of doubt and the desire to surrender.

While I strongly question every belief in individual genius or one person's "contribution" to a field of study, I do believe mistakes are the only things that one can – and must – *own*. I do both: humbly acknowledging the many immediate influences that have embossed this work *and* taking full responsibility for all its mistakes and shortcomings. They are entirely mine.

Scholarship is only one way of analyzing and discussing the world as it is, has been, might or might not be imagined. On a regular basis, I had the feeling that in view of other cultural products, my own work was banal and other people have found much better forms to articulate what I discuss. While the list of significant works and artists (past and contemporary) would be too ridiculously long – and any attempt of mine to create such a list as a kind of personal canon too ludicrous – I wish to name a selection here.

Coming from the theater, I am lucky to have had the chance to work with and for Lara-Sophie Milagro, Dela Dabulamanzi, and the entire Afro-German ensemble Label Noir. Their current production about Emmett Till will be groundbreaking for the German theater scene. Another shout-out goes to my friend and colleague Mbene Mwambene. His shows release the energy of affective frictions within various discursive forcefields as a multidimensional performative praxis with interactive storytelling, compelling poetry, unflinching sarcasm, music, and body movement. At one of his shows, I experienced how a carefully-staged 3-minute poetic vignette can do what I do in these 300 pages. I am so very pleased we met!

I also wish to acknowledge Spike Lee's *Bamboozled* for its impact on my work and person. The movie delineates the entire psychic reality of white people vis-à-vis Blackness in and as popular culture, both within the movie and in its reception. The scene in which Pierre Delacroix asks his writing staff to "tap into your white angst" probably influenced the way I thought about intuition and comedy long before I developed my analysis in this project. For all those who think this book is too long to read, I recommend watching *Bamboozled*.

Lastly and generally, I thank every Black-racialized person for every single moment they do not hold back on their critiques of whiteness when talking to me. For us white people it is literally unknowable how much energy that costs, and how much risk you run. This labor must never be taken for granted, and yet we rely on it. I recommend this attitude to every other white subject out there: you will not look or feel great, but you will look and feel less like a sluggish bag of psychic mess. Please always remember: us white people make the conversation difficult.

Contents

1	**Introduction: Spacing Whiteness in Improv** —— 1	
1.1	Lacking object and discipline —— 1	
1.2	Barthes's punctum as methodology —— 5	
1.3	White performance of Afro-pessimism —— 8	
1.4	Material —— 13	
1.5	Chapter survey —— 15	
2	**US Improv Comedy and Race – A Sketchy Report** —— 23	
2.1	Arresting improv —— 23	
2.2	Introduction to Chicago improv —— 25	
2.3	Talking race —— 34	
3	**Truths for Whiteness** —— 64	
3.1	Sylvia Wynter's conception of being hu/man —— 64	
3.2	The modern libidinal praxis of anti-Black abjection —— 76	
3.3	Embodying anti-Black abjection —— 91	
3.4	Modern popular culture as Blackness —— 111	
4	**Who Speaks?** —— 151	
4.1	Circulating aesthetics of vitalism —— 151	
4.2	Intuition and abjection —— 170	
5	**Abjection in Play** —— 186	
5.1	Performing Humanism —— 186	
5.2	Properties of play —— 188	
5.3	Improvising property —— 197	
6	**Funny Matter** —— 206	
6.1	Humorous Humanity —— 206	
6.2	Humor and the libidinal economy of anti-Black abjection —— 208	
6.3	Laughable Blackness —— 219	
6.4	Incongruities —— 228	
7	**In lieu of a Conclusion** —— 238	
7.1	Original ending —— 238	
7.2	Reality updates —— 239	
7.3	Retrospective reflections on scholarship in whiteface —— 244	

Works Cited —— 248

Index —— 259

1 Introduction: Spacing Whiteness in Improv

1.1 Lacking object and discipline

This project originated in the observation that the vast majority of people who practice improv comedy or improvised theater are white, while other improvisational or comedic art forms (jazz, freestyle rap, stand-up) are historically grounded in and marked as Black cultural production.[1] What is it about improv, which I know and practice, that makes it such a white space? But can an absence be an object of study – in this case Black absence? If so, what is there to study? Where should one look? And what would it mean to take improv as my object? Improv is a multifaceted and multidimensional phenomenon. When I use the term, I refer to its communal dimension (of a particular school, a local community, or the global improv community) as well as its aesthetics and aesthetic procedures. My usage encompasses how these play out in the theatrical situation of performance, its history and historiography, and the discourses that bring improv about and that it generates, such as manuals, articles, books, online discussions, and podcasts. I also denote its underlying Humanist assumptions, the performative role it plays within popular culture, and the way popular culture's social function feeds back into the cultural phenomenon of improv. One of the first questions that long stupefied me was how to disentangle improvisation as an aesthetic practice – more mode than content – from the ethical and political issues towards which Black absence points in loud silence. For too long, I adhered to the logic that improvisation was merely an aesthetic modality free of content that didn't mean in itself, and consequently didn't provide a ground for discussion outside homogenizing ascriptions to the absent group. Eventually, I came to realize that this colorblindness is part of the problem: Theatrical improvisation

---

[1] Throughout, I capitalize Black, Blackness, and terms that include these words – such as anti-Black racism – to mark their reference to a discursive construct rather than an empirical existence, which is not to say that those do not reciprocally affect each other. I refer to a signifier created by white people that, in my usage, has no ontological or essentialist content value and denotes no empirical referent. Again, this is not to say that those who are racialized as Black (in Nathaniel Coleman's formulation) are somehow indescribably unreal. The decision is a situated one: As a white scholar, I do not intend to make existential or ontological statements about what it might mean to live Blackness. Epistemologically, I am restricted to analyzing the creation and functioning of the signifier, not those signified. Black Studies scholars mostly use the lower case – and they are in the position to do so. As a white subject, I can make such existential (even ontological) statements only about whiteness or what white people do, so I use lower case for white or whiteness.

does not *see* color because it *is* color.[2] As a theatrical mode and in the history of its aesthetics and practices, improvisation *is* white, and continually sets up whiteness as a categorial default invisible to the white improviser. Thus, the question "Why are there so few Black improvisers?" becomes "How does improv continually constitute itself as a white space?" The ease with which improv discourse and practice provide a seemingly ahistorical grammar allows white improvisers to fancy their improv beyond the limitations of cultural and situated knowledge. Maintaining a focus on the subtle and not-so-subtle ways in which improv reasserts its whiteness is a challenge.

I am concerned in this project with the auto-institutive, or self-engendering, machinations of improv's whiteness understood as continual anti-Black abjection. The project is thus an attempt at white autocritique. However, where exactly is one to look in order to find the auto-institutive mechanics of improvisation? How does one lay bare the dynamics that underlie the superficial racial stereotypes performed on stage for everyone to see? What distinguishes improv from other forms of representational comedy or live theater? Previous scholarship on improv is of no help in answering these questions. Only one source deals explicitly with improv's dominance by white males: Amy Seham's 2001 *Whose improv is it anyway?* Seham's book has been ignored in academic discourse on improv and despised by the white majority of improvisers because it calls out racism and sexism. Most improv scholarship is written by improvisers themselves; it is always celebratory or seeks to demonstrate how improvisation and its principles can be applied in contexts other than the stage. Improvisation is written into theater historiography (see J. Coleman or Wasson), or used as a traveling concept at anybody's disposal; in theory, improv solves anything. Such theorization is aimed at legitimizing improv as an art form proper, designed to raise it to a level worthy of academic consideration simply by the act of considering it within the terms of the academy. This assertive approach is a problem for improv theory, which usually aligns concepts by analogy in one of two patterns of argumentation. One describes non-improv activities or behavior as improvisation or maps improv's poetic principles (such as the "yes, and" principle) onto non-aesthetic domains. The other applies academic concepts to improv, such as suggesting that in improv, we can see emergence at work. Both patterns are uncritically

---

[2] Aside from the titles of publications, I use italics to mark emphasis (as in this case) or to draw attention to a specific word in context. The latter sometimes involves semi-ironic nuances, not for humorous reasons but to highlight that the term a) is insufficient or inadequate for what is needed, similar to Derrida's erasure, b) reverberates with points previously discussed, or c) is polysemous and points out of its specific use and context toward something larger. I emphasize some words in quotations in the same way, and mark that emphasis.

in favor of improvisation and foreclose any critical treatment of the subject. In their attempt to generate universal knowledge, they rarely account for improv's less appealing realities.

However, neither its healing powers nor the application of improv's concepts to the non-improv world incited my inquiry. Disposing of these celebratory approaches, I looked at ways I could think about improvisation in less idealized terms around communicational practice, starting from familiar concepts of sending, receiving, gatekeeping, and so on. While I am still sympathetic to such close readings of an improv scene and of the modal aesthetics of improv, this approach soon proved to be too clinically objective; I found myself caught up in the intricacies of communicational concepts and the ill-fated ideal of methodological objectivity at all costs, which always derailed the analysis from my initial question. In every kind of allegedly objective methodology, I was missing something and I had a hunch it would never be enough. I realized that in using an aesthetic analysis that dealt with improvisation alone, I would never get to the core of improv's whiteness beyond the fact that white people always produce material based on and alluding to their shared white reality. That this reality is not universal is one of the central obstacles Black improvisers face in Chicago, as Aasia LaShay Bullock[3] and Dewayne Perkins[4] suggest:

> *Bullock:* It's so easy to imagine that the white experience is the neutral experience. That it's the experience everybody has. I actually didn't have that experience.
>
> *Perkins:* I was on a team in a theater. And they were talking about dad jokes, and we named our team "Daddy." They're like, "Oh, we should take pictures dressed up as fathers with Khakis!" And I was like, "My father doesn't dress like that. I don't know how to tell those jokes either."
>
> *Bullock:* That happens all the time. "Like mums do." And I'm like, "I don't know what that means. Explain to me. It never even occurred to you, because I'm sure all the mums you

---

3 Aasia LaShay Bullock is an LA-based comedian, writer, actress, singer, dancer, rapper, and music producer. Her web series *Starving Artists* won several awards at the 2017 New York Television Festival, among them "Best Short Form Digital Project." Most recently, she wrote for *Upload* (2020, Amazon Prime) and *Space Force* (2020, Netflix). Find out more on her personal website. Our conversation took place in March 2015 in Chicago, together with Dewayne Perkins. Bullock authorized the material used in this project via email on 6 October 2019.

4 Dewayne Perkins is an LA-based writer and comedian born and raised in Chicago. He won "Just For Laughs New Faces of 2019," and currently writes for NBC's *Brooklyn Nine-Nine*. He also wrote the Comedy Central web short *The Blackening* for the comedy collective 3peat, of which he is a member. *The Blackening* became a viral success with more than 15 million views. Find out more on his personal website. Our conversation took place in March 2015 in Chicago. Perkins authorized the material used in this project via email on 7 October 2019.

know did do that. None of the mums I know did. Guess what all your life is wrong. Let me shatter it a little bit. Let me tell you about my childhood." (personal conversation)

None of my go-to disciplines promised any significant results beyond the superficial critique of situated knowledges or culturally specific ignorance. Therefore, I decided to let go of disciplinary boundaries and trust my own (instinctive? intuitive?) knowledge about improvisation – as flawed and white-inflected as that must necessarily be.

Based on many years of improv practice and readings of improv literature, I have come to think of improv through three central concepts: a) intuition, because we do not have a lot of time to make decisions on the stage, and so rely on whatever helps us make them quickly, b) play, which is the primary descriptor of what we do, and c) humor, which is the most prominent outcome and aim of improv, at least in its US comedy version – despite some decidedly process-oriented artists and academic treatments that focus purely on its aesthetic dimension. Theories and practices of intuition, play, and humor all foreground some notion of the Human or involve statements about human existence.[5] They all feature some sort of celebration of the Human self. Recognizing this shared conceptual ground spurred my selection of the project's theoretical groundwork. The set of perspectives and concepts essential to this project includes Sylvia Wynter's elaboration of the modern West as a culturally specific construct that makes itself invisible to its cultural subjects; Sabine Broeck's concept of anti-Black abjection, which describes the continual dehumanizing anti-Black violence necessary for maintaining the white position as default; and Afro-pessimist points of critique put forward by Frank Wilderson. This deconstructive approach towards epoch-making Western Humanism, its repercussions, and the concepts used to describe them provides the theoretical and axiomatic ground for my three investigatory frames of intuition, play, and humor. Initially, I treated these frames separately, without any methodological guidance beyond grounding them in improv discourse and practice. However, in the process, I discovered that the language of psychoanalysis could be used to discuss what happens in improv in these respective fields. While they went in different directions, over the course of this project they became more and more interlaced, so that an unanticipated thread came to the fore. This project, then, is the conscious performance of *making* theory as opposed to applying concepts and presenting the results, proving or dis-

---

[5] I use the upper case for Human, Humanity, and Humanism when I refer to the cultural specificity of the modern subject, drawing attention to its status as a (semiotically arbitrary) discursive construct.

proving a thesis that may or may not have been part of a larger theoretical or political agenda. I bring into tense conversation previously unrelated or even disconnected theories and models of thought, keeping the argument always tethered to the grounds of an observed anti-Blackness in improv. This is original work in that it creates arguments about the relationship between anti-Blackness and improv that no single theory or allegedly objective methodology has yet provided or could provide.

This is not to say that I am not deeply indebted to previous scholarship as well as to those improvisers who have shared their perspectives and experiences with me. I am entirely dependent on critical race studies, theories of embodied cognition, psychoanalysis, and many more, as much as I rely on various manifestations of the improv discourse from without and from within. The latter includes a list of practicing (then) Chicago improvisers and comedians willing to share their (non-homogenous) thoughts, positions, and experiences with me, allowing me to use them here as well. The list includes renowned professional improvisers, young artists at the beginning of their careers, and everything in between. I have taken care never to analyze the interviewees or their contributions, but engage with them as interlocutors, cue-givers, and experts on the workings of the (Chicago) improv scene. They are never the object of analysis, but are rather on the team of analysts so to speak, even though I ultimately place their contributions dramaturgically in the logic of my argument. Nonetheless, I have taken great care not to present quotes out of context, instead maintaining their semantic and argumentative environment and retaining their autonomy as far as possible. This is why some quotations are rather lengthy. Where I did use only brief selections or turns of phrases, I checked to ensure I wasn't doing interpretive violence to their contribution. And yet, I am fully responsible for all shortcomings in this project, ethical and theoretical. As difficult as this terrain will prove to wander, I think this is the only way to talk about improv, given the lack of an object proper. The mere application of a closed theoretical framework, in terms of a mechanically consistent methodology, would bring a faulty order to the mess of reality and is thus predisposed to repeat the anti-Black acts of violence it aims to critique. Violence in the name of objectivity has done enough harm in the humanities.

1.2 Barthes's punctum as methodology

Given the absence of an object (or one with clear and observable borders), I develop a methodological procedure with recourse to Roland Barthes's notions of the studium and the punctum in his *Camera Lucida* (1981). Barthes sets out to develop a theory of photography, seeking an ontology of photography itself but soon discovering that photography as such is unclassifiable – as is improv. Its ontology cannot be reduced to separate, categorial units because there is no rule to determine the significant questions. Would they be empirical, rhetorical, aesthetic? In discussing the racial structures of improv, the same difficulties arise: do I consider institutions, empirical facts, rhetorical devices, sociological questions, tropes within the discourse, performance style, aesthetic foundations? What could the theoretical or disciplinary frame be? Should the work be semiological, sociological, psychoanalytical (8)? Leaving these questions aside, Barthes provides the pair of studium and punctum, which I will briefly introduce as a methodological model. The studium refers to the general meaning of a picture observed by a curious but rarely affected spectator. The punctum, in contrast, denotes a specific detail of the image that breaks through (*punctures*) the studium. The studium is an average and intended effect on the spectator, recipient, or reader, whereas the punctum is subjective, affective, and specific. The studium thus refers to what could also be called the hegemonic meaning in accordance with a dominant ideology, that is, the frame or discourse within which the image or discursive unit communicates and is shared by the assumed recipient as a quick decoder of its meaning. The studium gives us content-oriented encoded information.[6] The punctum, on the other hand, "shoots out of [the scene of the studium] like an arrow, and pierces me" (26). Punctum literally means "sting, speck, cut, little hole," making it "that accident which pricks me (but also bruises me, is poignant to me)" (26–27). The punctum is a detail that disturbs the studium, the mere presence of which "changes my reading" (42). What gives the punctum a kind of methodological impact and workability is its "power of expansion," which is "often metonymic" (45). The impact of the punctum is unintended, yet never accidental: "Hence, the detail which interests me is not, or at least is not strictly, intentional, and probably must not be so; it occurs in the field of the photographed thing [i.e., the analyzed object] like a supplement that is at once inevitable and delightful" (47). The punctum also speaks to an absence of the studium at large. Barthes claims that the "*punctum, then, is a kind of subtle beyond* – as if the image launched desire beyond what it permits us to see" (59). What narratives other than the studium (even running counter to it) can be read from the text when one departs from the singular punctum? What more extensive logic can be unraveled from this specific detail? What

---

[6] For further clarity: "To recognize the *studium* is inevitable to encounter the photographer's intention, to enter in harmony with them, to approve or disprove of them, but always to understand them, to argue them within myself, for culture (from which the studium derives) is a contract arrived at between creators and consumers" (Barthes 28).

makes it stand out (for me)? These punctum-focused readings will be applied as an investigative modality throughout this project, directed at elements immediately drawn from the world of improv as well as theoretical and ideological assumptions and concepts on which improv discourse rests.

In this study, I transpose these two concepts into the textual units of culture (culture-as-text) by engaging with the studium level to show the tropes around which improv positively writes itself into being, or to provide the background against which the punctums attain their effects. Sometimes it is a punctum or an absence that drives a close reading. Raising Barthes's dual model to the rank of a method also qualifies my project's position and its argumentational devices and strategies. Developing a methodology out of Barthes's concepts has implications. He acknowledges that the choice of analytical units and the allocation of their respective relevance to each other are always based on an affective/emotional "I like / I don't like: we all have our secret chart of tastes, distastes, and indifferences, don't we" (18)? The subjective evaluation of the scholarly I itself is very much central to any study (including this one). Barthes's approach is then to "remonstrate my moods; not to justify them; still less to fill the scene of the text with my individuality; but on the contrary, to offer, to extend this individuality to a science of the subject, a science whose name is of little importance to me" (18). He further writes that admitting one's subjective-affective punctums as examples at times involves "*giv[ing] myself up*" (43). Because a punctum is defined by a retrospective, emotional, and affective effect of the spectator or reader, it "shows no respect for morality or good taste: the *punctum* can be ill-bred" (43). These implications are valid for other scholarly enterprises as well, regardless of how objective, scientific, or methodologically accurate they fancy themselves. Such objectivity is part of the structure that veils and makes opaque the whiteness tacitly assumed to be a center of normalcy. With Barthes, we white scholars can carry our own positionality out into the open, without fancying ourselves in the position to disentangle our scholarly choices, the specific selection of our examples, and the general thrust of our argument. Even suggesting that I could put forward such an individual (not systemic) autoanalysis would undermine my central arguments.

The transposition from an emotional reaction to a text to its academic evaluation is purely heuristic. Thus, in a potentially devastating return to Barthes's foundational *I like / I don't* binary, this project is always vulnerable to criticism that positions it alongside other examples of white agents feeding on fetishized Blackness. Barthes himself is an example, even in the development of the studium-punctum concept.[7] And yet, because of our own white positional and affective grounds, how can we white scholars align our work with Afro-pessimist theory in a meaningful way, both in terms of strategy and with our own affective life? How can we put into words the intellectual reaction to and analysis of a real-world example or fictional story that generates disgust and repugnance but helps to develop, illustrate, illuminate a theory? Barthes has the same question, asking himself whether he could "retain an affective intentionality, a view of the object which was immediately steeped in desire, repulsion, nostalgia, euphoria" (21). How can we keep ourselves in check if we are not even reliably aware that we are performing anti-Blackness in our choice of examples or objects, in the racially ignorant application of inherently raced concepts or in our use of language? Why would the white scholar be in any way different from the white agent who is the subject of critique?

1.3 White performance of Afro-pessimism

Positionality and affect must be considered because Afro-pessimist performance – academic or other – takes on different meanings depending on the scholar who uses it. White thought in an Afro-pessimist framework is a paradoxical practice. We must ask whether the application (appropriation? obliteration?) of Afro-pessimist theory by white people can be something other than a transumed version of the "white desire to consume the spectacle of Black death" (Sharpe par. 100). Christina Sharpe addresses the different ways in which Black scholars and white scholars approach the theory, predetermined by their respective positionalities:

> I think that the only people who can *be* Afropessimists are non-Black people. I don't think Black people can actually *be* Afropessimists; my colleague, Kara Hunt, reminded me of this. We can theorize, we can meditate on Black suffering, we can experience the violence, we're marked. But we cannot *be* Afropessimists since the idea and reality of being is foreclosed to us: we're non-being. The only people who can be and embrace it are particularly these white, male, young academics who are so excited. They are *excited* by it. And it's an invigorating theory because it's a purely intellectual enterprise for them. (par. 81)

Sharpe speaks an uncomfortable truth for white male academics like myself. Merely pointing out this positional impasse does not solve the problem that a

---

[7] Consider Barthes's analysis of his affective reaction to "a family of American Blacks, photographed in 1926 by James Van der Zee" (43).

white scholar working with Afro-pessimist thought is in an ethically impossible position.[8] This paradox is only viable in the white supremacist system that provides me the privilege of speaking against my own position without *actually* annihilating myself, or better, where self-annihilation is a "purely intellectual enterprise." Is there any way in which this acknowledgment can be made constructive for the present project? Is asking this question part of the problem?

Firstly, the concepts and theories put forward by predominantly Black theorizers are not necessarily related to an ontological Black culture. Afro-pessimism is concerned with how Western society comes up with and engenders the concept of Black culture in the first place. My purpose is not to find out about Black culture, but to consider Western improv culture and its anti-Black abjectorship as its foundational and structuring principle. Theorizers of Afro-pessimism have analyzed these processes on an epistemic level, and it is here that white scholars need to learn from them, being willing students to "epistemic lessons in redress" (Broeck, *Abjection* 40). Theories put forward by Afro-pessimist thinkers are not ethnographic material provided by native informants and must not be treated as such. Secondly, as a white scholar, I do not intend to explicate or summarize Afro-pessimism but to listen to its thinkers and hopefully put its concepts to good use. The social position I inhabit precludes me from becoming an expert or attaining authority in the field. In writing and speaking from a white position on subjects related to white ignorance (as conceived of by Charles Mills), I am bound to make mistakes of different degrees and gravity. I am bound to trip over language all the time because no language is adequate to addressing the issues at hand, and if it were, it would not be comprehensible to me as a white subject. The language and knowledge I can mobilize is constitutive of the complex of problems I am setting out to discuss. Yet I believe it is better to engage in that messy discussion than do away with it in a mere nod of acknowledgment. Maybe there is a way to generate an argument out of white shame that does not end in issues of guilt or morality at large. Understanding the system in which I am positioned seems to be the best step towards building solidarity. This piece is therefore an attempt to be spoken-to by Black knowl-

---

8 Notions of ethics, possibility, and positionality are already deeply ingrained in the problematics I discuss. In many ways, such a statement is not true, because whiteness is the very structure that provides its subjects positionality for the obliteration of Black scholarship *without* any ethical queries or other sanctions. However, I leave the sentence accompanied by this footnote to show that the language available is itself imbricated with the epistemic violence under analysis. Any attempt to use it amounts to the deployment of the master's tools, to echo Audre Lorde. This is especially relevant when used by white scholars who fantasize about dismantling our own house.

edges and to listen, learn, and be vulnerable. To produce scholarship cautiously and with humility *in* and *as* practice. I do not invent, discover, or explore.

Positional specificity demands of the white scholar working with and using Afro-pessimism a constant alertness to the ways in which their writing as performative articulation are energized and inhibited by this configuration. This also relates to vulgar misreadings of such theorization as either entirely liberational and inspiring or as the epistemologically pessimist dead end it posits. At the same time, the white scholar must keep in mind how this alertness is in itself not particular but structurally libidinal. How do reporting on, foregrounding, and being the problem intersect? What is the performative effect of a white-positioned scholar who works with Afro-pessimist theory? In this context, it is important to note that as a white-bodied writing agent, I can (and should) say little to nothing about Black life, but I can say a lot of things about white structures of feeling that create Blackness as a signifier. While the sign is my creation, the referent is foreclosed to me. As a white-positioned author, I must be especially careful not simply to draw from an Afro-pessimist toolbox at will (will being a white privilege) and turn Afro-pessimist theory into a phrasebook. There is a genuine danger that terms, collocations, or phrases be applied carelessly and turned into hollow clichés. What does it mean for the white scholar to *look into the abyss* when they have no affective understanding of it because, for them, it does not exist? This is similar to the clichéd gesture of pointing out, which functions transitively to point certain elements of Afro-pessimist discourse and theory *out of the way* without actually reckoning with their consequences, or actively walking the theoretical space opened up by them. Pointing out denotes a gesture that allows the white scholar to speak without repercussion.

Further, Afro-pessimism demands a register incompatible with Humanist ways of reading and making the world, which involve concepts based on relationality and hope such as ethnicity, minority, diversity, and even difference writ large. In fact, in several sections of this project, this register becomes the object of critique. However, improv discourse manifests itself in these very registers. *Critical* positions regularly critique a lack of diversity or issues of representation. I do not wish to be the white scholar who comes along and applies a register that overrules the one applied by those who live the lives of Black improvisers in Chicago. Rather than whitesplaining the positions of my conversation partners, I draw on concepts located in the respective register of speech to address and value the material (opinions, anecdotes, and more). It is only in a second step that I move towards a different argumentational realm. What may appear a theoretical inconsistency in some sections of this project is my way of being spoken-to.

The impossibility of my knowing race-based dehumanization viscerally except from the abjector's point of experience confines the extent to which I can engage with it. However vague this may appear, I do believe that the concept of visceral, embodied knowledge is of crucial importance in the debate for Black authority on the subject. The knowledge necessary for such authority could never be in the bodies of white people. We just don't have it *in* us. The white body as a speaking subject-scholar does not exist outside the embodied anti-Blackness of whiteness I examine in this project. This is not an ethical but an epistemological issue. Moreover, the white scholar working in this field has to integrate a moral imperative (of engaging with Afro-pessimist theory beyond a mere nod of acknowledgment) with an epistemological impossibility – a situation for which dilemma is too small a word. The very attempt to integrate the imperative with the impossible is part of the specific modern epistemological project, which is built on the ideal of coherence (theoretical arguments, belief systems, subjectivities). Cherishing this ideal ignores the fact that in order to integrate both these positions, they must be imagined as existing on the same plane, even though the moral imperative has a completely different starting point and vector than the epistemological impossibility. Where the applied concepts perform and demand the destruction and collapse of the white author-subject, that subject paradoxically performs the maintenance of its own coherence in applying them. What then can a white person see *in the void?* What is the value of witnessing a cause that will not become a case? Specifically, how does the act of witnessing create the witnessed, and what libidinal economy and performance structure the relationship between the two? Is the impossible ethics behind this worth ignoring because of what might be won? (How) Can I speak about white ignorance without performing the claim for its transcendence? Is there any way for a white person to do this work without being – or thus becoming – an adventurer or a tourist? Is there any way to get aboard the ship and not be the captain, owner, investor, insurer? Is there, in fact, anything to be won? If so, for whom? Is white autocritique possible, and if so, what would it have to look like?

This project troubles the universalized, romantic concepts on which improv culture is based, leading to the complete dissolution of improv's referential and axiomatic framework without offering up anything in its place. I attempt to "learn how to go beyond ethnographic benevolence, as white European teachers, students, intellectuals, and how to practice disloyalty to white abjectorship and its ongoing power" (Broeck, "Legacies" 126). To do that, I mobilize Afro-pessimist concepts in a white discourse that needs to be troubled. Judgment calls about the performative or qualitative value of this project will vary. Do I risk too much for too little payoff? Can this project become part of an Afro-pessimist discourse, helping to generate a wider audience (in the improv sphere and in the disciplines I touch upon)? The individual white-positioned scholar cannot but become existentially brittle in becoming conscious of the fundamental antagonism of the modern world, working himself away and "wallowing in the contradictions" (Wilderson, "Wallowing"). Still, the aspiration to contribute something meaningful to Afro-pessimist discourse from a white position is not, or need not be, synonymous with an explicit or implicit call for white redemption or transcendence.

There is the cop-out of critical whiteness studies. But such a disciplinary label not only veils the fact that Black scholars have made this field possible; it is also structurally predisposed to re-center whiteness and tends to stay in the register of betterment, improvement, and progress. Moreover, moral judgment calls are often made in their name, or at least their discussions are grounded in moral implications, which makes it difficult to perform a structural analysis. I have taken care to avoid such gestures. This project is no fantasy of an antiracist transcendence of my own social position and racial privilege, nor does it mobilize the grammar of shame or guilt on the part of the subjects of my critique or for myself. It does not provide a method for so-called good anti-racist practice for white people, be it on the improv stage or in academic performance. I instead accept my structurally unethical position in the ethical conundrum, the epistemological mess of this politico-intellectual configuration, rather than pretending that this space is all tidied up – or could be via a critical academic performance.[9]

Building on the work and labor done in the field of Black studies always involves a double movement and contradictory performance for me. While in writing, speaking, or uttering I assert a social position of speakability that relies for

---

[9] Sara Ahmed's "Declarations of Whiteness: The Non-performativity of Antiracism" is imperative in many ways. She argues succinctly that "declarations of whiteness" are both common and foundational for these disciplines as well as non-performative: "they do not do what they say" (introductory paragraph). Her critique of the term "critical" in the disciplinary nomenclature is on point:

> I think the "critical" often functions as a place where we deposit our anxieties. We might assume that if we are doing critical whiteness studies, rather than whiteness studies, that we can protect ourselves from doing – or even being seen to do – the wrong kind of whiteness studies. But the word "critical" does not mean the elimination of risk, and nor should it become just a description of what we are doing over here, as opposed to them, over there. [...] The "critical" in "critical whiteness studies" cannot guarantee that it will have effects that are critical, in the sense of challenging relations of power that remain concealed as institutional norms or givens. Indeed, if the critical was used to describe the field, then we would become complicit with the transformation of education into an audit culture, into a culture that measures value through performance. (par. 8–10)

its existence on Black non-speakability, I also acknowledge the work of scholars and bodies of theorization under the umbrella term of Black studies. Black studies enables us to know and understand the universalist holes in the Human/ist discourse of improv and its theorization, and any attempt at white autocritique must necessarily start there. However, underneath all such ruminations lies the fact that from the white position, it is epistemologically and ethically impossible to adjudicate on these questions in the first place. It is not up to me to decide whether my work is or isn't part of Black studies, (critical) whiteness Studies, or any other studies, or whether it advances one or more political aims, as diverse as they may be (the notion of advancement being problematic in itself). I can never assess whether it is meaningful academic work *in solidarity* or whether it has a contrary effect.

1.4 Material

In what narrative or argumentational units and planes do I find punctums? What is the material of this project? My strategy is similar to the delineation of the investigative territories themselves and has had a similar trajectory. At first, I sought to define a clear-cut canon of improv: historiographies, manuals, and writing from within. However, this soon proved to be too restrictive, too methodological to get a well-rounded idea about what I wanted to discuss. Such a selection was liable merely to reiterate idealized improv poetics rather than provide a ground for analysis of what is actually happening. Next, I figured I needed to talk to people, to improvisers. An improviser myself, I thought there would soon be common ground and a collective endeavor to find things out. Three factors stood in the way of working with qualitative interviews as a method. First, analyzing, interpreting, or even critiquing Black interviewees would make me an ethnographer at best and an obliterator of Blackness at worst. Second, very few white people were interested in talking about the subject. Third, the conversations mainly took place in the register of multiculturalism, diversity, stereotypes, and so on, which communicates poorly with my theoretical framework. So just as I (intuitively) made out three fields for investigation, I chose to work with relevant material wherever it struck me. Sometimes a section from an interview inspired a conceptual or theoretical finding; sometimes it emphasized one. The conversations illustrate or provide counterpoints to my findings, which may or may not subsequently be deconstructed. I am aware that this makes the project vulnerable to a superficial critique in terms of method. However, I must emphasize that the well-trodden paths of transparent, mechanically objective, and blindly-applied methods should be challenged on these grounds because they ultimately satisfy only the critic and may have little connection to the real world.

The material of this project, then, is improv discourse in its broadest sense. It involves historiographies, encyclopedias, and manuals, but also podcasts, blogs, and other sources. I work with anecdotes from improv practice related to me by improvisers, situations and occurrences in shows I have seen or been in, and workshops in which I have participated. From a vast textual body on improv or improvisation, this material is subjectively selected to draw a picture of improv that will help to ask the central question: why is it so white? In analyzing this picture, I go back and forth between punctums and studia, because, as it turns out, in order to understand how and why a punctum is effective, it helps first to understand the ground rules of the studium underneath the initial superficial interpretation. In this project, punctums are generally elements within improv discourse and practice (past and present) that speak to or resonate with Black absence, anti-Blackness, and racism, puncturing improv's otherwise almost mythologically idealized existence. These punctums shoot through many planes of discourse and practice. In essence, this project is designed to draw lines between the planes. What is often viewed as a two-dimensional and chronological reality (a studium) presents itself in a multidimensional and transtemporal dot-to-dot drawing. Some dimensions are closer to the immediately observable, intelligible surface. Others are only vaguely pointed towards. Some are obvious from Black-positioned viewpoints but effectively inaccessible for whites, while others present themselves to us as whites like a far-away light, the source of which we cannot even begin to interpret. Some we can immediately recognize and address with the language at our disposal, while others may need more theorization, empirical findings, or the willingness and affective capacity for genuine autocritique.

Accordingly, I engage with these discursive units on various levels of abstraction. When working in and around the three (intuitively selected) conceptual pillars of improv discourse and their axiomatically-posited disciplinary ground – intuition, play, and humor – I engage with the ways in which they speak through and are made manifest in improv practice and discourse. I challenge the central tenets of improv – that is, the historical trajectories of phenomena or ideas – asking what exactly we understand by them, how they came into our world, and how else we might conceive of them. Only by questioning and challenging the meanings of these conceptual tenets can I get closer to improv's deep-seated linguistic and affective structures. As a result, the project often far exceeds the limits of improv, expanding beyond its existence as a manifest cultural phenomenon. While I set out to investigate the cultural phenomenon of improv, then, I also offer a discussion of the auto-institutive mechanisms of whiteness as articulated through improv.

1.5 Chapter survey

I begin my argument in Chapter 2, "US Improv Comedy and Race – A Sketchy Report," by laying out the general territory of investigation named improv. By improv, I mean not solely an aesthetic modality or stage practice but a cultural sphere that involves interrelated actual, social, and discursive spaces. For me, that discursive space includes everything from blogs to podcasts to printed how-to manuals and various almanacs. For this project, improv discourse denotes everything said about improv from within and without. I also differentiate this project from other scholarly engagements with improv, improvisation, improvised performance, and other variants or degrees of abstractification. I ground this project in the real-world thoughts, lived experiences, and perspectives of improvisers, who (at the time of the interviews) have mainly lived and worked in Chicago. I also draw examples from online debates about the racialized culture of improv. I have taken great care to keep the material grounded and connected to the actual improv reality.

First I will look at Chicago, demonstrating how this mecca of improv is an articulate manifestation of the way that social capital transforms into real career opportunities for comics. I mobilize the criticism advanced under the metaphor of a cult, which has often been directed against the improv scene. Then I move on to consider how race, and Black absence specifically, features in contemporary improv discourse, reading relevant discursive punctums as symptoms that are always overridden and brushed aside by the socioaesthetic ideals in which improv is embedded. Rather than marginalizing Black absence as an exception to the rule, I center this absence, looking at how improvisers today make sense of this absence and reading through the central topoi of this argument: concepts of (lack of) exposure, representation, economic factors, and so on. However, most of these arguments are presented in the registers of diversity or segregation – concepts about which those I interviewed held a vast range of nuanced positions. Throughout the years working on this project, I have come to adopt a different position on these questions. I present a critique of diversity via Jaye Austin Williams, Frank Wilderson, Karen and Barbara Fields, and Charles Mills, leading up to a consideration of Calvin Warren's theorization of a politics of hope. On this ground, the chapter ends with more examples of how anti-Black racism is a present absence in improv, arguing that there is indeed something specific in improv that makes it particularly liable to racialization and anti-Blackness.

Chapter 3, "Truths for Whiteness," presents a larger theoretical framework with which to approach the most fundamental points of reference of improvised theatrical play in improv practice, poetics, and discourse: the Human. Rather than idealizing this modern Human subject or accepting the romanticized axioms that regularly accompany the concept, I take up Sylvia Wynter's analysis of how the modern Human functions as a tool with which culturally specific Western Man "overrepresents itself as if it were the human itself" ("Unsettling" 260) while making this historically specific modality invisible to those who understand and experience themselves through it. Wynter shows that modernity does not bring about a discursive revolution of generalized human self-knowledge but rather a transumption of terms that still draws on a discursive matrix based on dichotomies of perfection and imperfection, heaven and earth, motion and stasis, high and low, reason and irrationality, Gods and beasts. Wynter traces this transumptive trajectory back to a Hellenistic "master code" on which, as she argues, modern whiteness has been modeled as a human ideal ("Unsettling" 263). Within this code, Blackness functions as the sphere of white Humanity's absolutized other. I follow Wynter in her conceptualization of being human as a praxis in *bios-mythoi* hybridity. Taking my cue from what she develops as the sociogenic principle after Fanon, I relate her theory to the disciplinary frameworks of psychoanalysis and embodied cognition. I understand both to be instrumental in Wynter's overall aim to conceive of being human as praxis rather than as noun. The significance of Wynter's array of concepts for this project cannot be overemphasized – most centrally the notion of transumption (borrowed from Harold Bloom), but also the sociogenic principle, being human as praxis, her in-depth analysis of the discursive matrix, and the alternative pathways she opens up to understand what being human in bios-mythoi hybridity may entail. While I will occasionally mark the most immediate connections, Wynter's work has been instructive in many regards for this entire endeavor. From here, I turn to Frank Wilderson's notion of libidinal economy and the symbolic and affective role of anti-Black enslavement for the making of the white modern subject's psyche. I consider more explicitly those theorists associated with Afro-pessimism and emphasize the libidinal dimensions of the modern subject, whose coherence I argue is entirely dependent upon discursive and affective anti-Blackness, the social death of Black-racialized people, and the longue durée or "afterlife of slavery," a term coined by Saidiya Hartman (*Mother* 6).

Drawing on Wynter's dictum that "[w]ith being human everything is praxis" ("Catastrophe" 34), I introduce the concept of anti-Black abjection as theorized by Broeck in acknowledgment of but distinct from Julia Kristeva's concept of the "abject." In this project, I think Kristeva's and Broeck's usages of the term abjection alongside each other. Kristeva develops her concept within film theory, and I suggest that the affects engendered by *perceptive* abjection are important to understand, or at least to keep in mind when reading specifically anti-Black abjection as a *continual praxis* of being human in Western modernity, as Broeck does. With Broeck and Kristeva, I develop the historically and culturally specific notion of an embodied modern subject-effect that must be continually engendered to maintain individual, social, and political coherence. In Western modernity, this subject-effect is structurally always already achieved by anti-Black abjection qua action, perception, and imagination of fungible Blackness that is open for gratuitous violence of all kinds. In a section of Chapter 3 titled "Embodying anti-Black abjection," I look into how anti-Black libidinality and the idea of a subject-effect can be discussed in the language of embodied cognition and neuroscience. Drawing mainly on Antonio Damasio's classic and still highly influential concepts, I consider emotional, feelings, and somatic markers. I understand the linkage between psychoanalysis and embodied cognition as putting into practice Sylvia Wynter's model of being human in bios-mythoi/logos hybridity. I use both psychoanalytic and neuroscientific concepts to argue for the cultural specificity of the discursive (and affectively biologized/naturalized) structure of Western modernity, leading up to reading Blackness as a somatic marker – both specific and concrete as well as vague, undefined, and potentially expansive – through which anti-Black abjection happens. Linking this back to Kristeva's notion of a subject-effect, which I reframe as a subject-*ae*ffect, I theorize white solipsist subjectivity in Black presence and absence as a discursive and biological predisposition for voluntary or involuntary anti-Black thought, affect, and action. At the end of this subsection stands a metatheoretical consideration of Kristeva's abjection as an already-raced concept. I engage with the argument that the application of abjection runs the risk of circular reasoning: the concept seems so apt for analyzing racism because it is itself a raced concept, starting from the condition it seeks to critique. I preempt this criticism by returning to the decidedly (and exclusively) anti-Black abjection of the modern West understood as an act rather than a feeling, as well as to Wynter's conception of being human in bios-mythoi hybridity and her post-Fanon theorization of the "study of the Word" as a scholarly method. Unlike Kristeva, I do not aspire to conceive of or engage with a fixed ontology of human being.

In the next section of the chapter, "Modern Popular Culture as Blackness," I lay out the field of popular culture understood as a disciplined territory in which improv exists. However, before looking into improv as such, I put the popular itself under scrutiny by reading it through the lens of modernity's anti-Black abjective structure. After a brief consideration of the ways in which the rhetoric of appropriation is insufficient when it comes to anti-Blackness, I historicize the birth of US popular culture (and, by extension, popular culture in the modern West) in the spectacle of the auction block. I draw mainly on Saidiya Hartman's invaluable archival work in *Scenes of Subjection*. This discussion also further defines the fine line between white witnesses to and spectators of dehumanized Black suffering: what Hartman discusses as the "terror of pleasure" and the "pleasure of terror" (*Scenes* 32), and the white subject's incapacity for empathy when confronted with the suffering of Black bodies. I look at the way lynching parties functioned as popular cultural events and galvanized the creation of a popular as such. I go on to consider James Baldwin's short story "Going to Meet the Man" as an in-depth analysis of how the "pleasure of terror" experienced by white subjects through anti-Black abjection is central in the development of US popular culture. As Baldwin's story contains graphic descriptions of dehumanizing violence against Black people, the analysis is preceded by a trigger warning. A subsequent discussion of minstrelsy creates a more obvious connection between my project and what is more commonly known as popular culture. In minstrelsy, we find an overt articulation of anti-Black abjective culture as well as the aesthetics of stage performance in a way that white readers will find more recognizable. In this section, I mainly draw on Mel Watkins's extensive monograph *On the Real Side* (1999); unlike its subtitle suggests, the book is not (only) *A History of African American Comedy*, but also provides ample material for understanding US popular culture at large.

I conclude the section by looking at how popular culture has been treated in contemporary scholarship, arguing that proponents like Richard Shusterman draw on the same affective and matrical assumption as popular culture's classic opponent, Theodor Adorno. The discussion here takes the shape of a close reading of the latter's infamous criticism of jazz (once again, some might say) within the critical framework of Afro-pessimism and the conceptual toolkit developed through this project. The underlying hypothesis is that Adorno can make his generalized arguments about popular culture only by way of abjecting and – in the strict sense – dehumanizing Blackness. Grounded in this analysis, I suggest that all popular culture is both Black and anti-Black, that the popular is and has always been a sphere for white, sociopolitical negotiation that substantially references and works with Black cultural production but never addresses this very axiomatic assumption. Taking my cue from and in conversation with Fumi Okiji's *Jazz as Critique*, I address Adorno's arguments about jazz outside the moral or judgmental register while acknowledging the racism and anti-Blackness necessary to understand the full dimension of his argument.[10] In a coda, I argue that Adorno's positions on popular culture are dated in the sense that we had better take popular culture as a fact rather than a romanticized (or scorned) space of political negotiation. I bring Adorno into conversation with Ronald Judy, who discredits such defenses of modern, individualist subjectivity as regressive. Seeing the difference between Adorno's holding on to modern Human subjectivity and Judy's gestures towards a different, affect-based idea of human being altogether opens up significant conceptual pathways.

In the next three chapters, I work around what I have defined as the foundational concepts in improv practice and poetics. Chapter 4, "Who Speaks?," engages with the idea of intuition. I begin by reading the aesthetic and philosophical climate of improv's emergence, which cultural theorist Daniel Belgrad has defined as an epoch and aesthetic configuration named the "culture of spontaneity." In willful ignorance of potential friction between the sphere of the popular and the assumed sphere of art, I lay out how the axiomatic similarities that underly those post-World War II artistic practices and the popular, humorous world of improv are very much the same; both seek to vivify an uncritical conception the Human in a conceptually esoteric but sincerely applied neo-vitalism. With recourse to Donna V. Jones's *The Racial Discourses of Life Philosophy* (2012), I demonstrate how vitalism has been inherently racial from its inception. Its turn-of-the-century and early twentieth century manifestations in particular – most notably its articulation by Henri Bergson – were the ground for the circulating aesthetics in which improv could flourish. These neo-vitalisms worked overtly on anti-Blackness for theorizing and propagating the invigoration of white but universalized subjectivity at a time when human beings were understood as automatons. Given that anti-Blackness is the formative ground for this neo-vitalism, anti-Black abjection should be visible in both art and popular culture. I show how this proves true not only in aesthetic performative practice but also in its theorization, demonstrating how the very Blackness that gave life to, inspired, and enabled philosophies, poetics, and practices has been effaced. While the Black-as-Slave figure has been mobilized as a central metaphor for what the white subject does not want to be, Blackness itself has been obliterated. The historical contexts of Black cultural production have been willfully thrust into oblivion, and its principles are turned into an all-encompassing American-

---

10 I hold Toni Morrison's assertion about the literary imagination to be just as valid for the scholarly one: "A criticism that needs to insist that literature is [...] 'race-free' risks lobotomizing that literature, and diminishes both the art and the artist" (12).

ness. By close reading a passage from Sam Wasson's *Improv Nation* (2017), I demonstrate how improv discourse does this explicitly.

In the second section of the chapter, "Intuition and Abjection," I zoom in on the notion of intuition. The concept is mobilized by improv and within the broader culture of spontaneity to gain access to something higher (truth or design), or, from another perspective, to something lower (the unconscious). I consider Viola Spolin's classic writing on intuition in *Improvisation for the Theater* (1963) as well as another seminal improv text, *Truth in Comedy* (1994), which represents Del Close's approach, showing how two of the most influential teachers of improv draw on the concept of intuition. Drawing on Damasio, Wynter, Mills, and Warren, I then make a case for understanding intuition not so much as a pathway for an unconscious potential resigning within *us*, but as a mode that allows white people to generate a subject-aeffect – the linguistically arbitrary but biochemically real auto-imagination of ourselves as complete, integral entities – and which by necessity draws on anti-Blackness. I thus argue for the cultural specificity of a white intuition as opposed to a generalized human intuition assumed to bring *us all* together, partaking in a Politics of a Truth we are falsely assumed to share as a universal. Towards the end of the chapter, I return to the modern libidinal structure, speculating about whether the fact that people achieve a subject-aeffect qua culturally specific anti-Black abjection might more radically reveal that for us white people, intuition as such is anti-Black. I ask whether it is reasonable to distinguish intuition as such from intuition qua Blackness as a contingent, arbitrary signifier. That we can generate a subject-aeffect through intuition in the first instance (not the second, *cultured* one) is also based on the fact that we white people can only know ourselves as a split person inhabiting the world of reason and the world of beasts. For us modern white subjects, this very split can only exist in anti-Blackness; if not for anti-Blackness, then, intuition would have another affective effect for us. It would not necessarily make us feel whole. It might not even make us feel anything at all.

Chapter 5, "Abjection in Play," deals with the second discursive pillar of improv: play. I draw on D. W. Winnicott's theory of play and the attached concepts of transitional objects or phenomena. With Winnicott, I define play as a space in which the fictional and the real, the affective and the symbolic are activated simultaneously, making it a highly fitting concept for analyzing improv practice in general. For the infant, the objects or phenomena at the center of abjection move back and forth between these two spheres. Winnicott describes these objects as the first *properties* of the child, and their movement back and forth between the real and the fictional is part of its subject formation. They are transitional in the sense that they perform a central function in the transition from union with the mother into the symbolic realm. Activating the sphere of symbolic non-existence and the exit from non-symbolic bliss in playing is "essentially satisfying [...] even when it leads to a high degree of anxiety" (Winnicott 70). At this juncture, I link Winnicott's theory of play to Kristeva's understanding of abjection. Kristeva too builds on the notion of a child rejecting a mother's breast, which for Winnicott functions as a (symbolic and real) referent for which the transitional object or phenomenon has a (symbolic and real) stand-in function. Winnicott's theory of play is insightful in two ways. Most importantly, the transitional object or phenomenon shares central characteristics with Blackness. It provides the child with an experience of omnipotent control both over the object and its subject state. Through transitional phenomena, the child generates and rehearses its (primal) subject-aeffect. (Winnicott does not restrict this to physical objects, even though these are his primary concern. He also considers transitional phenomena, by which he means songs, motions, and other forms of play. Very much like the mobilization of intuition, the notion of Black transitional phenomena allows us to think about the activity of play in the same way we can think of a white agent's affect when listening to R'n'B or watching interracial porn.) Additionally, Winnicott's transitional object or phenomenon can be related to the anti-Black abjectorship of white subjectivity because it is always property *for* the child.

In improv discourse, the phrase "to treat somebody like a prop onstage" is common. I offer a close reading of the human body as a theatrical prop in a section titled "Improvising property." What does it mean to have a prop on stage? How can a human body be made to function as a prop in a scene? How does anti-Blackness feed into this aesthetic and metaphoric propertization of human bodies on the stage? After establishing the conceptual simultaneity of the literal and metaphorical understandings of a human body as a theater prop, I consider in more detail functional and communicational theories of property from Margaret J. Radin and Carol M. Rose. As implied in the German term for prop – *Requisite*, which comes from Latin and roughly means requirement or necessary thing – I ponder how the human-as-prop(erty) is a necessity for (white) improvisation based on a Western subject that requires possession and power over things and bodies to maintain itself.

Chapter 6, "Funny Matter," addresses the most obvious pillar that structures improv discourse: humor. On the grounds of the libidinally anti-Black structure of the modern West, I consider Freud's traditional analysis of joke work, the modus operandi of which he compares to dreamwork, and its specific mechanics of jouissance. I focus on what he calls the tendentiousness of jokes, the libidinal undercurrent that, according to Freud, drives *all* humor – regardless of how abstract or aesthetic it may fashion itself. I link this tendentiousness, its communicative functions, and the way it engenders sociality (both temporary and beyond time) to the anti-Black abjective principle. I suggest that the specific tendentious power of humor in social situations predetermines anti-Blackness, at least whenever Blackness comes up in a staged and performed situation. Afterwards, I enter a metatheoretical register and consider incipient continental humor theory. I demonstrate how anti-Blackness has metaphorically and non-metaphorically engendered the theorization of humor most explicitly and with it the theorization of the Human – the a priori referent (and logical end) of all humor studies. In Bergson, who surprisingly reappears in this discussion, we find an almost textbook anti-Black humor theory on the grounds of his universalist, neo-vitalist humanism, as discussed in Chapter 4. While with Freud I cover the so-called release theory of humor, with Bergson I engage with what has come to be classified as a superiority theory of humor. I argue strongly that in theorizing humor, there is no justifiable legitimation for such differentiation, extending this argument to the allegedly abstract and aesthetic approaches of so-called incongruity theories, which are the ground of most contemporary humor studies. I trace part of the historical trajectory of this strand of humor theory and discuss how it has always been intertwined with other theorizations of humor, arguing that there is no reason to distinguish them from one another – even if only for heuristic reasons. I further suggest that, paradoxically, such heuristic distinctions mystify the fact that in effect and in its historical trajectory, the ground for such incongruity has always been the foundational antagonism of the white Human subject and the de-Humanized Black body. This discussion of humor is summed up in a section titled "'Where is your brain from?' Blackness-as-superpower." What might appear as a final turn toward an optimistic Afrofuturism at the end of the project is rather a consideration of the ambivalent, contradictory, and hard-to-bear humorous powers that the foundational difference between Humanity and non-Humanity holds *for those positioned as Human*. The Blackness-as-superpower topos came up in numerous conversations with improvisers in Chicago, and involves all the messy, contradictory, hopeful, and pessimistic positions and realities that have driven this project. Following the previous theoretical analyses, this section re-grounds the project in the realities of improv comedy. It stages the ways in which being Black in improv for some of those who live this discursive impossibility "is almost like a handicap that you have to make a superpower. If you're the only black person in a scene, you're automatically driving that scene. It's not fair, but it's true" (Joel Boyd, personal conversation).[11]

---

[11] Joel Boyd is an improviser, writer, and stand-up artist. He wrote for *Drop the Mic* (2018, TBS/TNT) and *Earth to Ned* (2020, Disney+). He also created, directed, and wrote for the web series *Sad-Ass Black Folk*. Boyd now lives in LA. Our conversation took place in March 2015 in Chicago. The material used in this project was authorized by Boyd via email on 15 November 2019.

2 US Improv Comedy and Race – A Sketchy Report

2.1 Arresting improv

Improvisation is ephemeral, transient, and cannot be captured. What is true for improv as a special kind of stage performance is also true for its academic analysis. Yet, improv or improvisational theater, is gaining momentum in academic consideration. Scholars analyze poetics, write chronological histories of its aesthetic development, delve into fantasies of orality, truthfulness, and humanity, or re-discover allegedly lost didactics. In most studies, theatrical (or other) improvisation serves as a means of discussing something else: concepts of emergence, models for creativity, antidotes for depression, effective team-building, non-authoritarian creativity, artificial intelligence, and humanity as such. The number of TEDx talks that mobilize a core feature of improv practice – the "yes, and" principle – is steadily on the rise. Improvisation is a hot topic, and we practitioners usually claim we've always known that. However, the mere application of various principles of improvisation to different fields does not necessarily tell us more about improv itself, and very few scholars work on improv in its theatrical variant. Gunter Lösel is one of the few who attempts to figure out disciplinary and definable frames for talking about improv – for example, in his differentiation between improv as an autonomous art form or a cultural impulse ("Phänomen Impro"). Both frames can help us see improv, and both offer an interesting take on the practice. Nevertheless, Lösel's underlying assumption that artistic expression and cultural impulses are separate spheres of cultural activity is, in my opinion, dated and grounded in an implicit hierarchy between high and low popular culture. Such approaches thus amount to little more than the accumulation of works that academize improv with little payoff. When scholars approach improv, then, we do not necessarily mean the same thing. We assume that we share ideas or interests – such as creativity, emergence, and collectivity – but that is not always the case. We assume we are talking about the same thing in different languages or disciplines, but we forget that changing the frame changes the thing. We must understand that scholarship is *always* interested, and that its object is constructed by the methodological design, by the selection of material, by the doxastic environment and methodological procedures. The scholar before their object is Goodman's spectator before a painting: just as there is no virgin eye in art reception, there is no objective method. Instead, all we can do is a) describe what we're drawing on to create the object in the first place, and b) reflect on and present our own theoretical and methodological axioms. The former is the subject of this chapter, and the latter will be considered in the following one.

This project understands improv as a theatrical stage practice (aestheticized play) based on spontaneous decision-making (intuition) practiced in local and global communities of the modern West, mostly aiming to be comedic (humor). These three planes, though structured in three chapters, shine through the argument at all times and must be understood in their simultaneity and transareality. Where other theorizations of improvisation work with improvisational concepts concerning something else, I develop an analysis of improv's libidinal economy. This project aims to feed into the theory of improv as a stage performance – not the other way around. Further, I focus on the realities of what actually happens rather than the allegedly endless potential of improvisation, and on questions of racism and anti-Blackness as they speak through theatrical improvisation. Improv is neither solely an aesthetic modality nor merely a stage practice. It is also a cultural space, which I differentiate into actual social spaces – people who practice and perform improv – and discursive space, which is generated by and feeds back into the social spaces. This discursive space encompasses everything said about improv from within and without, from blogs to podcasts to printed how-to manuals and various almanacs of improv, as well as the positions that non-improvisers take towards it. I draw no significant line between academic and practical discourses on improv because both fall into self-idealization. This project is not concerned with affirmative or reproductive historiography or celebrations of self-proclaimed political idealism as performed through a theory of absolute aestheticism. Although I am by no means an opponent of theory (as the reader will soon discover), I ground my analysis in real-world performance practices. I write with actual stage performances in mind, or discussions with people who have told me about their stage experiences – and not improvisation's presumed "capacity to trouble the assumptions [...] fostered by dominant systems of knowledge production" (Caines and Heble 3). The position that imagines improv as a space for radical oppositionality and social improvement is held by many improvisers. When asked how improv compares to other art forms with regard to discriminatory practices and exclusion, renowned improviser TJ Jagodowski[1] says:

---

[1] TJ Jagodowski began improvising in the mid-1990s. He has learned, taught, performed, or directed in all the improvisational theaters of Chicago. He has been at iO Chicago for twenty years, at the Annoyance for fifteen, and wrote and performed in two Second City reviews. His commercial, TV, and film credits include the Sonic ads, *Stranger than Fiction*, *Ice Harvest*, *The Great and Powerful Oz*, *Prison Break*, and *Get Hard*. He can also be seen in the 2018 web series *Studio B*. Together with his partner David Pasquesi, TJ has performed the fully improvised "TJ & Dave"

Even though I think improvisation was slow to come to greater diversity, I think it holds great, if not the greatest, ability to deal with it. Improvisation allows the performer the most freedom of any art form to speak their minds directly and without editor. To speak from their direct experience most momentarily. But it is also inhibited in affecting change by its audience size and impermanence. A book or film can reach millions and is a permanent work. It can be studied and peeled apart. It can be disseminated widely. The improv performance is often to a group of 20–100 and vanishes into thin air. The tenets can be spread wide, but again that's through missionary work by teaching 10–20 students at a time. (Jagodowski, personal email, 16 Oct. 2019)

While the improviser in me agrees in part with the position that improv has the potential to deal with discrimination in theory, the theorist in me is curious about this assumption's paradoxical relationship with reality. I hope that my approach, located precisely in that paradox, will prove productive and go beyond the mainstream reiterations that make up the bulk of academic improv theorization.

2.2 Introduction to Chicago improv

I am working within American Studies and therefore focus on the type of improv developed and institutionalized in the US. Yet, my three planes of analysis (intuition, play, humor) and my axiomatic framework (a theory of modern anti-Blackness focusing on its libidinal dimension with psychoanalytic models) could speak with no loss of significance to the several European traditions of improvisation. Improv has been inserted discursively into various historiographic strands and different trajectories, and European styles have developed without conscious reference to the improv practices in the US. They often reflect artistic approaches distinct from comedic entertainment and do not relate as much to the comedy industry as US improv does. However, much of what I am consider-

---

show on many stages throughout the USA, Canada, and Europe. The show has won numerous awards, such as the Del Close award two years in a row (2003 and 2004), and the Nightlife Award New York (2006 and 2007). Both performers have been celebrated as the "Chicago Improviser(s) of the Year" (2006) and Chicago Reader named "TJ and Dave" as the "Best improvised Show" in 2008. Among critics, colleagues, and audiences, Jagodowski is widely considered as one of the best improvisers currently performing. See further their duo website and their book, *Improvisation at the Speed of Life* (with Pam Victor). Our conversation took place in March 2015 in Chicago, together with David Pasquesi. The authorized material used in this project is drawn from a follow-up email exchange in October 2019.

ing here also applies to these stage practices, and one needs to start somewhere – even if the object of study is as boundless as improv.

Geographically, I focus on Chicago, historically the institutional source for US improv culture as it is known today. For many, Chicago has always been the place to go for training in the comedy entertainment industry. Amy Seham calls the city "a mecca for young improvisers who want to study it at its source" (xvii). The Chicago improv scene remains an active sphere where people are trained and socialized to enter comedic entertainment at large. With Second City "as a stepping-stone to Saturday Night Live and other opportunities in film and television" as its preliminary endgame (Seham xvii), we can read Chicago as a hotbed for the development and cultivation of comedic talent, and the space in which this talent gets unpaid stage time and opportunity to present itself:

> When I decided I really wanted to pursue comedy, I feel like Chicago is the kind of city where, wherever you come from, whatever small town you were at, you're trying to hit the ceiling there. And if you could make it in Chicago, I feel like it's a good stepping stone to either go east or west or even London or whatever. (Joel Boyd, personal conversation)

Based on the number of students, performance frequency, historical interconnections, and success in terms of popular fame, three Chicago improv institutions stand out: Second City, iO (formerly ImprovOlympic), and The Annoyance Theatre. These three are referenced most often in the interviews, and represent different focal points of central aspects in improv: the communal, the satirical, and the dirty. The schools have distinct aesthetics and provide different frames and styles for training and performance. They generate different audience expectations, identification points, and frames for self-perception. Claims for distinction from other schools are made along similar lines. Such distinctions are most pronounced in published manuals, histories, encyclopedias, and almanacs: the Second City has published the *Almanac of Improvisation* (Libera 2004), the iO is represented in the seminal *Truth in Comedy* (Close et al. 1994) and later *Art by Committee* (Halpern 2006), while Mick Napier describes the Annoyance style in *Improvise. Scene from the Inside out* (2004) and *Behind the Scenes* (2015). Each institution advertises its training centers and classes through the names of alumni who have made it, such as Tina Fey, Keegan-Michael Key, Stephen Colbert, Dan Castellaneta, Scott Adsit, Rachel Dratch, Amy Poehler, Tim Meadows, Andy Richter, and many more. Credibility and authority on all things improv and comedy are further marked by the school-specific manuals, historiographies, and their preface-writers: Mike Myers vouches for *Truth in Comedy,* Adam McKay believes in *Art by Committee,* and Bob Odenkirk cherishes Mick Napier's

Behind the Scenes. Second City's *Almanac of Improvisation* (2004) features texts by Anne Libera, Tim Kazurinsky, Tina Fey, and many others. However, assigning these public figures to individual schools or institutions is arbitrary: Mike Myers is also an alumnus of Second City, Bob Odenkirk performed at the Second City as well, and so on. Actors who have gone on to great fame can be as easily associated with the iO as with Second City, like John Belushi or Bill Murray.

One institution outside Chicago needs to be mentioned here: *The Upright Citizens Brigade Theatre* (UCB) in New York City. Founders Matt Besser, Matt Walsh, Ian Roberts, and Amy Poehler were students at the iO in Chicago before they went to New York and set up the UCB. Its success was enormous, and UCB has become another hallmark of Western, US-based improv. It too has a manual: *The Upright Citizens Brigade Comedy Improvisation Manual* (2013). Their official training website announces that UCB alumni have gone on to become successful "writers and performers for *Saturday Night Live, The Tonight Show with Jimmy Fallon, Broad City, Key & Peele, Silicon Valley, Veep, Atlanta, Brooklyn Nine-Nine, Crazy Ex-Girlfriend, The Daily Show, Inside Amy Schumer, Master of None, Drunk History, Full Frontal with Samantha Bee*" and others. The UCB's training program has earned it a central role in the wider entertainment industry; their website calls it "the only accredited improv and sketch comedy school in the country." Improv training is common amongst comedic performers on TV, and in casting for films, improvisers are often given preferred treatment over actors who graduate from traditional theatre or drama schools. TV and streaming productions increasingly feature improvisers: series like *Curb Your Enthusiasm* or *30Rock* draw primarily on improvisation and cast people specifically trained in it. The cast of Saturday Night Live (SNL), the endgame for comedic performers, regularly features improv-trained members.

While training in improv and active participation in the scene may eventually lead to jobs in the industry, in the early stages it demands time, money, and a willingness to network and socialize. Stage improvisation rarely pays, and hardly anybody can live only on performing improvised shows. What improv offers to its young practitioners is stage time, performance practice, and the training to produce funny material quickly for all kinds of comedy. Participating in improv as such is an expensive endeavor and a social and financial privilege:

> Bullock: The amount of jobs that improvisers can get here that pay money is a small percentage in comparison to the amount of improvisers there are. So it still seems like most time when you're auditioning, it's not for anything you know is going to pay. A lot of times, it's just for exposure, to be on a team – and hopefully, a popular team that people will see, and then maybe they will let you work for free for a while, and then maybe they might hire you.

> *Perkins:* And a lot of times that you audition, you have to pay for it. You have to pay for the coach. There's not a lot of financial gain.
>
> *Bullock:* It's a lot of investment. I had to spend so much money before I ever started making money doing improv.
>
> *Perkins:* There are things like internships at certain places. But then you have to invest so much time. That is a lot to give to this art until you are able to be paid for it.
>
> *Bullock:* And even then there's not enough jobs for everyone who works hard and pays all this money. It's not a thing you do to make money. It's a thing you do to build skills that will help you make money. (personal conversation)

This social and symbolic capital can outweigh the cost of classes. Improv also provides a social space for people to get together, a large community of improvisers who are happy to practice and perform together, who share their worlds on- and off-stage. Social relationships also provide access to stage opportunities:

> The community is active. It works so much on getting to know people. That's how you get stage time. These smaller theaters, they don't pay. You get to know people; you make teams; it has to be social. There are a few places in which you can audition and put on a team, but for the most part, teams are independently made just through people you kinda like. (Bullock, personal conversation)

> I've been doing theater since I was a kid, fourteen years old. Improv pretty consistently since 2001. My father met Charna, and he was like, "My homegirl got a theater, it's called the iO, and you'll do good there." That's where I started. I was one of the first interns for the TJ&Dave show. Charna was nice enough to let me take the classes for free because of my dad. (Johnson, personal conversation)[2]

Chicago improv is an intricate entanglement of social life, practical training, and opportunities to start building a career in the entertainment industry. There is a strong link between social capital – having a lot of friends and the kind of humor

---

2 Warren Phynix Johnson graduated from improv programs at iO Chicago, ComedySportz Chicago, and Second City. His stage credits include "Pimprov," The Second City's National Touring Company "RedCo," Second City revue "The Second City Does Baltimore," Crossbreed Comedy with Laugh Factory's house team, "Bastards Of The Underground," "Lady Mechanics," and "Awful People Present: Stealing From The King Unabridged" (both at The Annoyance Theater). Johnson has hosted Chicago Public Schools' own television series *CPS RIGHT NOW*. He has performed in *Improv Unlimited* (corporate improv shows) and was an ensemble member in McDonald's Corp's only traveling improv show, "America's Next Success Story." He has featured in several net commercials (Comcast SportsNet and McDonald's), and broadcasts weekly on B.A.D. Waves on Que4Radio (que4.org/itunes Radio) with the Bastards Of The Underground. Our conversation took place in March 2015 in Chicago, together with Dacey Arashiba and Derek Schleelein. The material used in this project was authorized by Johnson via Facebook Messenger on 7 October 2019.

to which a majority can relate – and career opportunities. Being able to get into improv requires financial investment or personal connections. Without financial capacity and social connections, the joys of improv are inaccessible to many (including potential access to or success in the comedic industry).

Some describe improv's exclusive dynamics, the power of social capital, and the heavy investment in belonging as cultish, usually with a strongly negative connotation. Stand-up comedian Peter-john Byrnes works the cult metaphor in a 2017 article for the *Chicago Reader* titled "Why improv is neither funny nor entertaining, according to a stand-up comedian:"

> Improv is to comedy as Scientology is to religion: it suckers white people into paying ever more expensive fees to the organization to gain higher levels of achievement. Like any cult, its hierarchies are of endless fascination to those within it and deadly boring to those without. And like Scientology, improv is centered around a messianic leader, though the example of the late Del Close suggests that L. Ron Hubbard would have been even more toxic if he'd had a smack habit.[3]

The semi-ironic religious connotations of being part of the iO "cult" are articulated not only by those denouncing it, but are fostered by the theater itself. The iO mainstage is named after the late Del Close, whose role in improv cannot be denied: as a teacher, performer, and mentor, he directly influenced much of Second City's history, and was the artistic mind behind the foundation of the iO and the development of the Harold as the most-performed improv form and iO signature piece. Charna Halpern, owner of the iO and Close's long-time collaborator, has cherished the rumor that Del Close's ashes and skull sit on an altar at the iO, and never tires of relating the mythology of his life and death (Butler 35–36, Halpern, *Art* 99–130).[4] While is not part of my project to discuss how to judge Close's character, it helps to consider briefly the discursive function of his figure for improv's self-making discourse. Close was an important contributor to what we know as US improv, even "one of the most influential figures in

---

[3] Byrnes is neither a much-noted critic nor a comedian, but his article sums up some arguments made against improv in general. Regardless of his intent or qualification to speak on the subject – his curtailed understanding of the "yes, and" principle is disproven in many other examples considered – he puts forward a view on improv that is real, whether or not it is "true." Also, aspects of his article resonate with arguments put forward in the discourse, like the cult or religion simile considered here. Moreover, reactions to the article from within the community – by professionals and non-professionals, most notably on Reddit and in the comment section of the article – speak to how insiders perceive themselves and how improv is perceived by non-improvisers.

[4] Another account of his death is given in Kim H. Johnson's "As Del Lay Dying" (2003).

the history of American comedy," as Eric Spitznagel writes in "Follow the Fear," a balanced essay on the person and his mythologization. Close has attained a carefully-developed and maintained status as the most important one, the only one, the guru. He fabricated a thorough mysticist and anecdotal narrative about his own life, which he celebrated as a mixture of truth and things that sound like it. Spitznagel, when asked to write a profile of Close for a Chicago newspaper after Close's death, was confronted with many secondhand accounts from colleagues and friends, but "nobody seems to know with any certainty whether any of his stories are true." Spitznagel observes that "since his death in 1999, Close's stories have become a permanent fixture in improv mythology." The discourse is not short of publications invigorating this mythologization. Spitznagel summarizes:

> There've been two films, including PBS's *The Legend of Del Close* (2000) and *The Delmonic Interviews*, a feature-length documentary that still screens on the national festival circuit, most recently at the Phoenix Improv Festival in April 2007. "Bring Me the Head of Del Close," a stage show that consisted solely of performers telling stories about Close, played to sold-out crowds at the Strawdog Theatre in Chicago during late 2004. Jeff Griggs wrote a memoir, *Guru: My Days with Del Close*, about his experiences working as Close's assistant and errand boy, and Kim Johnson – who, along with Close and Charna Halpern, cowrote the quintessential textbook on improvisation, *Truth in Comedy* – is reportedly working on a more comprehensive biography. Most famously, there's *Wasteland*, the short-lived DC comic book that Close cowrote and considered his autobiography.

While imagined (or remembered) as an outstanding and influential comic, as an eccentric guru of sorts who mixed theater and improvisation with drug use and celebrated the Gonzo lifestyle, Close was much a child *of* his time, rather than *beyond* it. Both improv and Close's role in it must be read in their historico-cultural context, as Seham does in her historiography of what she calls the second wave of improv:

> New Age religions, televangelism and fundamentalist religious sects, and "self-religionist" or self-actualization movements [...] and Scientology emerged to fill the empty space of any unifying or collective belief system for many Americans in the 80s. (33)

Close claimed to have been friends with L. Ron Hubbard, to whom he had allegedly suggested that if he needed money, he should "just turn Scientology into a religion," only to state later that "Just because it didn't happen doesn't mean it isn't true" (Spitznagel). The language of religion and spirituality found a direct way into Close's improv practice and discourse. In the improv bible *Truth in Comedy*, the reader learns about a performance game or an exercise called "invocation," in which "students invoke a 'god'" (Close 108–09). What may sound like a metaphor, an image to give students to develop something from nothing, may at the time have been taken literally. Improv students may have thought they were in fact invoking demons, and Close may have believed himself to be "protecting" them against those demons (Halpern, *Art* 104).⁵ And after the suicide of his mentee James Belushi, Close sought help in spiritual leadership:

> Rather than turn to counseling, Close found renewed inspiration in unlikely places. He immersed himself in the occult teachings of author Aleister Crowley and became the Warlock at a local Wiccan coven. Close was so enamored by his new spiritual beliefs that he began drawing on Wiccan rituals for his improv workshops at the Second City. (Spitznagel)

Through figures like Close and the stories associated with him, we learn about the structure and discursive mechanics of a community that creates and maintains this figure. There is something to be learned about a community in the figures around which they are built, in this case, a messianic leader and authoritarian gatekeeper of the "real" knowledge of universal truth, humanity, and humor. This also involves the exclusion of those unwilling or unable to get initiated (and accepted) into the cult, and, like other cults, manipulates through the language and ascription of talent (or lack thereof). If we can assume that some people *have it*, then we are quick to believe that those uninterested in such a cultish community do not. The idea of talent in combination with authoritarian assessment is, in fact, cultish. If Del Close had been a Black woman, improv would not be such a white space now – but then it might not even exist, because the discursive space of improv could never provide a Black woman figure with the unquestioned respect for the messianic leadership ascribed to Close. Close has been discursively chosen, collectively appointed, perhaps even retrospectively created as a guru – regardless of who he actually was or

---

5 To the contemporary reader, this might be far-fetched speculation. But we should keep in mind that it was the time of New Age fetishes, which are regularly raced, as we see in the following quotation. The idea of improvisers actually invoking demons is part of the mythologizing discourse as reified by Halpern, who recalls her first encounter with Del Close:
> Art the performance, he gathered his students in a circle and began to do something called "The Invocation." And, since it was Halloween, they decided to invoke demons. At the time, I was taking classes in meditation and had been taught to do something called "white lighting" yourself. It was a method of protecting yourself by imagining your body surround in white light [...] After the very creepy performance, I decided to go up to him and give him a piece of advice of my mind [...] I said to him, "You had a lot of nerve invoking demons without protecting the audience." He responded in a condescending manner, "I protected the building." "You can't do that," I responded. He stared me down and said, "Yes, I can." (103–04)

what he actually did, about which I can make no reliable claims. But so much can be said: understood as a structural space, improv has little issue with such cultish guruism.

However, some parts of the community do react when the paradox becomes too obvious, as happened in a controversy about an annual improv marathon in New York at the UCB in 2017. The marathon was initially advertised with a poster that Seth Simons describes and interprets for Paste Magazine as follows:

> Let's just mull over the image for a moment. So, that's Del Close right there with the big face and glasses. He appears to be embedded in the side of a mountain, Mount Rushmore-style. Or maybe he's lying against it, the rest of his gargantuan body out of frame. There's the Statue of Liberty atop his head, for some reason, and a neanderthal at the bottom, gazing quizzically upward for some other reason. Then we have everyone either ascending to the top or celebrating there, men and women, and some guy wearing salmon-colored shorts. And then there are the faceless masses approaching in the darkness. And then, to the side, two words: "The Wokening." What is The Wokening? Is it Del Close? Is it the Marathon? Is Close literally saying "The Wokening," and it's, like, a speech bubble, perhaps signified by what looks like a megaphone to his right? Has he just shouted "The Wokening" and everyone's streaming toward him so they don't miss The Wokening, which reads sort of like it may be a reference to "The Happening," or maybe it's just its own thing and not a reference at all? Unclear.[6]

To present Close in this way as the leader of an improv community (local, global, transreal) caused a backlash mostly among non-white improvisers on Twitter and Facebook. This punctum in the discourse sheds light on the space in which my project is located. Whatever truths can be said about Close, as an improv community, we should be wary of the gurus we choose and what they say about us. A culture that indulges in such a personality cult, including the modalities and contents that come with this specific figure, will necessarily be uninteresting or off-putting for people looking at it from the outside – or example, in Byrnes's general derision of improv, or in this observation from Kimberly Michelle Vaughn,[7] who simply does not play along with the celebration of Close:

---

6 The poster is reproduced in the article by Simons.
7 Kimberly Michelle Vaughn graduated from Columbia College of Chicago with her BA in Theatre. An alumnus of The Second City Touring Company ("Red Co"), she opened "She the People: Girlfriends' Guide to Sisters Doing It For Themselves," Second City's first show created and performed by women. She also performed in Second City's "#DATEME," "Legendary Laughs," and their 107th Mainstage Revue "Algorithm Nation or The Static Quo." Her TV credentials include *The Chi* (Showtime) and *Written Off* (Amazon). More information is available on her website. Our conversation took place in March 2015 in Chicago. Vaughn authorized the material used in this project via email on 8 October 2019.

> Everyone in that community hails Del Close. They love him. That man did not believe that black people can improvise. He did not think women can improvise. I don't give a fuck about him. (personal conversation)

UCB's official reply to the backlash was largely judged appropriate. Still, the undercurrent was defensive: "He was a provocateur, anti-establishmentarian and someone that truly cared about the misfits and weirdos," and, "While Del's contributions cannot be overrated what we are attempting to highlight are other voices empowered by this art form" (Simons). In these comments, we can see part of the problem. This issue does not lie in the fact that his contribution is somehow underappreciated. Rather, it lies with the fact that its racial axioms, its violence, and its anti-Blackness are unrecognized. While these features sometimes run along with discussions of Close as a guru figure, they are not read as their logical consequence. Toni Morrison asks us to be "mindful of the places where imagination sabotages itself, locks its own gates, pollutes its vision" (xi). This is also true for understanding improv and its original leadership by Close.

So is the idea of a cult helpful for this project? Is there more to it than the recognition that improv provides for its community and philosophy (Noble), or that "most popular improv advice sounds like spiritual challenges" (Hines)?[8] First of all, understanding the discursive role of improv's contemporary practices of self-description, specifically the iO, is crucial. Second, the idea of an exclusive

---

8 Also compare the reaction to Byrnes's article by improviser Nelson Velazquez, who, in a blog post, agrees that
> Improv IS a Cult (Sort Of). There's no shame in saying this. [...] Many people love the feeling improv brings of personal freedom to act like school children or finally finding like-minded, passionate individuals that love to do some things on stage or maybe just having a break from their mundane life for a few hours a week as an escape gets them on the hook. Those feelings and interactions can be addictive. You want to feel like that all of the time so continue to do improv classes, attend shows, hang out with all kinds of people, and do your own shows. That's called community and every cult's got one.

Velazquez does not accept, however, the notion of spiritual leadership for the entirety of the improv world:
> The article goes wrong in not realizing that many of us don't "center ourselves around a messianic leader." There are many people across the world that are very much admired for their ability to use improv to make us laugh, cry, get angry, or feel uncomfortable on a consistent basis. Like any artform, we have our "heroes" for sure but in improv our heroes encourage us to find our own identities as actors and avoid being like them. This is completely contradictory to a true cult where leadership wants you to conform to what they preach irrespective of what you want to do with their teaching and will take actively take steps to shut down any form of dissent that may threaten their position. Improvisers look to inspire their own work from the greats out there. This is where the cult analogy breaks down a bit.

cult with an uncritically accepted belief system applies as well. However, in this context, this belief system is partly bound up with but not restricted to the cult around a person (guru) or church (school), but needs to be understood in a deeper and wider context. The belief system is not the Harold as a gateway to ultimate truth. It is not Del Close, Second City, or the iO. The cult is called humanism, and its belief system is whiteness.

2.3 Talking race

Black punctums

In a culture that claims universal truths and humanity, Black absence – as well as its symbolic hypervisibility in the last couple of years – has been one of the most relevant, telling, and silenced talking points in improv discourse. If universal truths, human relatability, community-building, and even social healing are striven for through humor, and where satire is argued to provide marginalized voices the opportunity to speak, Black absence constitutes a problem. If egalitarian play and a general anyone-can-do-it-attitude are promoted, Black absence speaks to something for which improv discourse has no register, no grammar, no vocabulary. A community that understands itself mostly as liberal, egalitarian, welcoming, inclusive, can only do so by ignoring every voice that indicates a contradictory reality. Roger Bowen, an early producer of improvised theater, gives insightful commentary about why improv is such a white space. When asked "Why are there so few Blacks in improvisation?," his reply serves as a paradigmatic argument for understanding the anti-Black elitism of improvisers' self-conception as satirists:

> I think that satiric improvisational theater is definitely a cosmopolitan phenomenon and the people who do it and its audience are cosmopolitan people who are sufficiently liberated from their ethnic backgrounds to identify with whatever is going on throughout the world. They know what a Chinese poem is like and what Italian food tastes like. But I don't think most black people are cosmopolitan. I think they're more ethnic in their orientation, so when they're black actors, they want to do black theater. (Bowen qtd. in Sweet 40)

The concept of cosmopolitanism can only function here semantically in its difference from what Bowen makes out as an "ethnic background" within a racialized binary of possible conditions. The white cosmopolitan state of being is synonymous with the political subject of the modern West, which Sylvia Wynter identifies as "Man1" and will be considered in more detail in the next chapter. This state involves the imagined transition from a natural to a civilized state of being by way of "sufficient" liberation from an "ethnic" state of being. This liberation, according to Bowen, finds expression in a refined taste and perfected human development. It is characterized by universality as such, which involves spatiotemporal and cultural omnivoracity. "Ethnicity" or "Blackness," on the other hand, is the background from which the modern, cosmopolitan man is distinguished. Bowen argues that improv is too intellectual for those racialized as Black. In this way, Black-racialized skin forecloses the possibility of a Black improviser by signifying their incapacity to liberate themselves. Bowen adds:

> You see, ethnic art tends to emphasize, enhance, and reinforce certain states of values, to say, "Our group is a good group." But when you get out of that and you identify with a larger intellectual environment, you say, "Well, gee, that was pretty narrow stuff." You get a concept of the brotherhood of man and how much alike people are rather than how different they are. You become de-ethnicized and you become a citizen of the world. [...] Blacks aren't at that point. The ethnic experience is very enjoyable, but it excludes the outer world. It's always "Us against them." In some ways it makes it easier for a person to get along because he doesn't have to fight every single battle. Now, a cosmopolitan has to fight every single battle there is because he can't say, "Me and my tribe say, 'Fuck you,'" because he has no tribe anymore. The cosmopolitan person also, by the way, is in a position of having to improvise a whole way of life, whereas in an ethnic society, much of it is handed down to you; it's received tradition. (Bowen qtd. in Sweet 41)

Here Bowen juggles ahistorical concepts, cultural descriptors, and social positions, positioning himself and others in whatever way he sees fit. His unpersuasive arguments and lofty semantics answer the perceived lack *of Blackness* posited in the question. Bowen's is a paradigmatic example of how the repertoire of white defensive arguments can be put to use to minimize or mask the uncomfortable truth of improv's obliterative aesthetics. Even though the original question focuses on an absence, an irritation, a punctum, a lack *in improv*, Bowen is capable of explaining – even legitimizing – this absence by relocating the lack within Blackness rather than in the white space of improv where it was originally observed.

Over the years this project has taken, Black absence from improv has erupted several times in the discourse. My starting point was a then-recent 2013 article in the *Chicago Tribune*, in which Meredith Rodriguez engages with "the lack of diversity" on Saturday Night Live. Rodriguez traces the absence of Black cast members back to the improv scene of Chicago – and its racialized configurations – as the major breeding ground for future comedians. She observes that "those leading Chicago's improv and comedy scene say that although women and minorities have been breaking down barriers in the last several years, finding enough minorities for [Chicago's Improv] main stages remains a challenge."

She is seconded by Andrew Alexander, then CEO of Second City, who agrees that "[w]e always have to do more," adding that "[t]he bench is fairly thin" and "[i]t's not like we have a lot to draw on." Alexander seems to be doing the best he can, investing "millions of dollars in Second City's diversity program" over the last two decades, financing "workshops at inner-city schools, casts that feature minority talent and scholarships at the Second City training center." According to the article, Second City's Outreach and Diversity coordinator Dionna Griffin-Irons had an annual budget of $200,000 at her disposal in 2013. Charna Halpern, the head of iO, faced similar problems despite advocating a race-neutral perspective: "Halpern does not keep quotas or look specifically for minority actors, she said, saying that she simply looks for the 'best players.'" Halpern continues: "I have some coming up for further auditions. [...] They just need to get wet behind the ears." In this article, Black absence is presented as something to be solved and measures have been taken to do so, with slow effects. The same day Rodriguez's article was published, *The New York Times* featured an article titled "'S.N.L.' to Add Black Female Performer" by Bill Carter. The article reiterates the allegedly race-neutral argument by stating that executive producer Lorne Michaels is "purely driven by talent considerations." This alleged colorblindness is undermined by Michaels himself, who states that it is "100 percent good for the show to have an African-American woman." Carter points to Michaels's claim that this "is not merely because the show could use a woman capable of playing the first lady, Michelle Obama, in sketches," but reminds the reader that this is "important enough a consideration that all the candidates will be asked to try an impression of her Monday night." Notably, this decision was not arrived at without public and individual pressure:

> "S.N.L." had been subjected to a barrage of criticism over the last several months over what seemed to be a glaring absence on the comedy show, which has had relatively few black female performers over its long history. The criticism was kicked off by comments from two of the show's black cast members, Jay Pharoah, who said the show needed to "follow up" on the promise to add a black woman, and Kenan Thompson, who announced he did not want to do any more impressions of black women in drag. (Carter)

With the introduction of UCB-trained Sasheer Zamata to the SNL cast in 2014 (she stayed until 2017), the debate was assuaged until September 2015 when a long and detailed blog entry by Oliver Chinyere re-opened it. In "Why I'm Quitting UCB, And Its Problem With Diversity," Chinyere states:

> UCB does not care about black people or minorities. It does, has done and will continue to do the bare minimum when it comes to maintaining diversity not unlike the entertainment industry at-large. As nine openings on house teams quietly came and went, not one POC

[Person of Color] was added, despite the fact that in the past year, two POC have stepped down. We are technically less diverse from a racial standpoint.

Chinyere's blog entry was widely received, and such temporary interventions on Facebook or in online newspapers and magazines function as powerful punctums within and against the white studia of improv. Keisha Zollar, former diversity coordinator at UCB, wrote a much-discussed post on Facebook in response to Chinyere, and a 92-comment Reddit thread continued the conversation ("r/Improv"). McDonald takes up the matter in an article for *The Washington Post* titled "Diversity problems persist for Upright Citizens Brigade comedy troupe, students say." She interviews Chinyere:

> This isn't the first time these issues have bubbled up in the comedy troupe. Chinyere told The Post that a group of students met with the UCB "powers that be" in January 2014 and offered suggestions for how to improve the experiences of students of color. "Nothing was addressed on that list," she said.

Chinyere is not the only one to address the issue online. Other examples include the blog *Miss Adventures of Milly* (mirrytamalez, "My two cents about the lack of diversity in upper levels of UCB classes"), an interview with Peter Kim who also chose to drop out of UCB ("Comedian Talks Quitting 'Dream Job' At Second City Due To Racist Audiences"), the follow-up to Kim's article by improviser Simon Tran ("Resisting Racism on the Improv Stage"), Patrick Rowland's "What to expect if you're a Black Improvisor," or Keisha Zollar's appearance on the *Acting Income* podcast hosted by Ben Hauck, in which she discusses unconscious bias poignantly and in rich detail.

There are also rare instances of white engagement with Black absence, such as the episode "Inside the Master Class: Black Guy Auditions" in a UCB Comedy Youtube series about diversity. Rarely do these go beyond semi-ironic white helplessness in articulating anything on matters of race, as in the similarly sarcastic 2012 *Onion* article "Everyone In Improv Troupe Balding." If there is a discussion on anti-Blackness – most commonly in the register of "diversity" – it is always grounded in the presence of a Black racialized performer and never the subject of a white-on-white conversation. This is not a silent position per se, but improv discourse in general is not only extremely slow to react; it also finds no way of understanding how its *fundamental whiteness causes Black absence*. It never looks into itself. Ignorance has traditionally been and remains its ultimate silencing strategy. Those who talk about it are almost always Black performers.[9]

---

9 While the above examples relate to UCB New York (except for Chicago improviser Patrick

In 2015 I conducted several interviews with improvisers, and the response to my call for interviews affirmed this; none of the improvisers to whom I was directed or who responded to my wide-reaching call were white. While I am happy that improv stars TJ & Dave (TJ Jagodowski and David Pasquesi) were willing to talk to me, that contact was based on a personal encounter. It was always automatically understood that a) I wanted to talk only to Black people, or b) it is a Black people's problem of no concern for white people. This lack of white interest on questions of racialization comes as no surprise to the Black-racialized interviewees who were willing to talk to me. All Black improvisers addressed structural racism and told personal anecdotes that revealed improv's racism to be as incredibly present and powerful as the ignorance it encounters in white improv culture at large. Joel Boyd describes this phenomenon with an impressive effort at empathy:

> A lot of issues where race comes up in comedy or improv in Chicago, there is a very small tight-knit group of people who discuss those things between each other. I only talk about it because most of my friends do that. I don't know that a regular Joe Shmoe improviser who moved here from Charlotte or somewhere, who just came here to do improv, I don't know, what that person is thinking about.
>
> In general, a lot of white people who wouldn't necessarily encounter these kinds of problems on a day to day basis – How would they seek it out? It's a problem they don't have. It's a problem that they don't understand. And that they don't need to understand because they live in that other world where it doesn't exist for them, so why would they? I mean, I am sure that that's what's going through their head. It's not even racist. It's just– Their lives are fine. If this issue was solved or wasn't solved – their lives would be the same. I understand why white people don't talk about it on a daily basis. It's untalkable. I do it on stage because I think it's very important for us to not be silent about this kind of stuff. It's not gonna get solved if you don't say anything. There are a lot of deep social issues today that probably won't get solved just because people aren't willing to talk about it. But at least improv can bring those things to the light so that people at least talk about how they felt about this one scene after the show.
>
> Whenever I have conversations about stuff like this, I always feel like it has to be at the right time and place. It almost feels like you have to know what you're getting into. The people getting into that conversation have to know that they are coming into the room to have that conversation. The only times I have talked about this have been forums and discussions among black student union. Even our outreach and diversity at Second City have sponsored two or three events with pizza and stuff, and they bring people to talk about stuff like this. But other than that, in our society, people just aren't ready to talk about it. It's such a "hush-hush put it under the table." I think if that was an aspect of im-

---

Rowland), the issue is as Western and global as anti-Blackness itself. From London, Tai Campbell writes: "Black people don't exist in improv. Well, there are about five of us. In a world where talking unicorns are real and time travel is possible, we don't exist. We don't exist as characters (even offstage), as performers, teachers, or as prominent voices in the improv world."

prov classes, it would help. At least a section, like a day: "This week we're gonna focus on this." I definitely think you couldn't do it as a weekly thing. Nobody would come back. But I definitely think that it's a road, that if we don't take it, it's going to get worse. (personal conversation)

Patrick Rowland points out the paradox of a "welcoming" community unable to or uninterested in realizing improv's assumed potential to "be for everyone," zooming in on the need for pushing the debate:

> This community is crazy welcoming. They welcome you, but it's weird that it's that thing where it's so few of us. And I'm not only talking about black people. I'm talking about Latinos; I'm talking about Asians, they're even a smaller percentage than black people. Any team at Annoyance or anywhere and you'll see it's just a lot of white guys, maybe one or two women, and maybe somebody of color. I think the reason that me and Nnamdi [Ngwe] are all more willing to talk about this is we're trying to get that word out, trying to get more of us to take classes, so everyone knows it's not just for white people. Not just for the tall, lanky white guy with glasses. It's for everybody. And I think we all need to work harder just spreading that message. I think because people are so used to it, they forget that there's this whole other margin of the world that is missing out on it because you're not talking about it more. (personal conversation)

Kimberly Michelle Vaughn is not surprised over the lack of interest from white improvisers:

> I don't feel like that's on their mind. I don't think that's their goal or their mission. They won't say, "Let's get some real voices in here. Other voices than ours." I am not saying that their voices don't matter, but still: "Sit down, we've heard yours before. What the fuck are you complaining about?" I honestly don't know what we can do to change that. (personal conversation)

Dacey Arashiba[10] wonders if white improvisers may simply feel that they have nothing to say about it – again making Black absence an issue for those absent rather than an impulse for white autocritique. He also considers the effect that an overwhelmingly white community performs with the lack of people who understand racism, and who are able and willing to call it out:

---

10 Dacey Arashiba writes and performs sketch comedy with the Asian American comedy ensemble Stir Friday Night! (see ensemble website), and sings in a punk rock choir called "The Blue Ribbon Glee Club." He was a professional musician, and one of his compositions features in the movie *D2: The Mighty Ducks 2*. The conversation took place in March 2015 in Chicago together with Warren Phynix Johnson and Derek Schleelein. The material used in this project was authorized by Arashiba via Facebook Messenger on 7 October 2019.

> The white kids might think they don't have anything to add to the conversation. There is no canonical opinion about race in improv. I've seen plenty of things that devolved into stupid racism. Because there are no people of color in the crowd, nobody is paying attention to how stupid some things actually sound. Sometimes you have to call kids out about this: "It's a cheap laugh, but you gotta approach it differently. How do you play this smarter?" (personal conversation)

If it was structurally understood that the absence of Blackness means something for improv as a whole, a sensibility might emerge for the community, the cult, at large. But as a cult, improv has its teachings in place, and lacks the teachers who could address it adequately:

> [Racism in scenes] is not talked about in class, really. Most teachers won't call it out. There's a thin line between stifling the student – 'cause they're paying for this class – but also you don't want that at the expense of your other students who are not comfortable. So it's a weird line for teachers to walk and I understand that. But if you don't have an advocate when you first start, if you don't have a teacher who has your back, someone who can tell you that it's not just you, I wouldn't believe that it gets better. It does get better. Some teachers will talk about it when it comes up organically, but it is not part of a curriculum. I wish it was. How to be a good teammate and how to not make your women castmates or castmates of color feel like props on stage. That should be something that's taught. And most of the time, when I tell people how I feel about what they do and what I find offensive, it's the first time they're hearing it. So a lot of people of color just ignore it, because it's hard to always have to be the person who says "Can't do this. Don't do this. That's wrong. That's mean. That's offensive." You don't want to always be that person because you got a reputation. But if you don't, a lot of people won't know. But how would they? A lot of teachers don't teach or even only address it in class, which makes it not a fun learning environment. (Bullock, personal conversation)

Racializing factors

Those who do acknowledge the reality of Black absence regularly relate the racial question to other forms of discrimination and debates about the representation of "minorities." When speculating about the reasons for Black absence in improv, three explanatory fields come up repeatedly in the discourse: a lack of exposure, argued within the logic of representation; issues in infrastructure that center around geographic location and culture; and socioeconomic factors, including a culturally specific libidinal economy. First, every improviser I talked to identified a lack of representation and general awareness of improv in Black communities as one of the main reasons for Black absence:

I think when theatrical improvisation started, it was a white person's game. Why that is, I don't know, but for a long time, there were no or very few minority performers that a young minority person could look to in order to emulate. I worked in the box office at Second City, and during my downtime, I would look at old cast pictures and can count on one hand the minority performers that were in those pictures for the first 30 or 40 years of the Chicago theater's existence. I don't know the Toronto or Detroit casts at all. And the same can be said for old pictures at ImprovOlympic. So, without seeing themselves represented in the histories of those theaters, I can see it not seeming like a place where they felt welcomed. It feels like improvisation was slower to arrive at its black stars than say, stand up or music or film, but I only know my corner of the improvisational world, and improvisation was slower to find its stars in general. It feels like only in the last twenty years that people have understood nationally what improv is and who its star pupils were, with the occasional exceptions of people like Nichols and May. (Jagodowski, email follow-up to personal conversation, 16 Oct. 2019)

I think the reason that the vast majority of improvisers are white is that the vast majority of people exposed to and interested in improvisation are white. Or rather have been. I also think that is changing. During the time I have been involved with improvisation, I have seen women better represented as each year passes and the same with other groups of people. I believe that as the pool of improvisers better represents different groups, improvisation becomes more available to potential new improvisers. Current improvisers are the ambassadors and recruiters for new improvisers. Some people need to see someone who looks more like themselves doing it in order to say, "Hey, they're like me. Maybe I can do it too." (Pasquesi, email follow-up to personal conversation, 20 Nov. 2019)[11]

Black absence from white improv is understood as a kind of magical vicious circle. To recognize the lack of representation is not to understand its reasons, and suggesting that Black absence is the cause of Black absence is circular logic. What other reasons are mentioned?

---

11 David Pasquesi has been improvising since the early 1980s. He studied with Del Close for years and was on one of the original Harold teams at ImprovOlympic (now iO). Pasquesi wrote and performed four reviews at The Second City, where he received a Joseph Jefferson Award for Best Actor in a Review. He has played Steppenwolf and the Goodman theaters, and his many film and TV credits include *VEEP, Strangers with Candy, Boss, Angels and Demons,* and *The Ice Harvest.* Together with his partner TJ Jagodowski, Pasquesi has performed the fully improvised "TJ & Dave" show on many stages throughout the USA, Canada, and Europe. The show has won numerous awards, such as the Del Close award two years in a row (2003 and 2004), Nightlife Award New York (2006 and 2007). Both performers have been celebrated as the "Chicago Improviser(s) of the Year" (2006) and the *Chicago Reader* named "TJ & Dave" the "Best improvised Show" in 2008. For more, see his website, the "TJ & Dave" website, and his book (with TJ Jagodowski and Pam Victor) *Improvisation at the Speed of Life.* The conversation took place in March 2015 in Chicago, together with TJ Jagodowski. The authorized material used in this project is drawn from a follow-up email exchange in October 2019.

The argumentational foundation of Black absence from Chicago improv is the city's racial segregation, and the parallel distribution of financial resources. The three main Chicago improv theaters are all located in areas with a low to non-existent Black population. (See Alana Semuels's article in *The Atlantic* for a contemporary analysis of "Chicago's Awful Divide.") The Second City and The Annoyance are in Lake View; iO is located in nearby Lincoln Park.

> I am teaching several high schools in the Southside. I teach improv to high school students. And they weren't aware of Second City or SNL. Audiences in the Chicago area, especially downtown, in the Northside, because that's where most of the art is, and are all white. You barely get an occasion of a minority in there. It's great when you do, but you want more. We want to commercialize more and advertise more out there. There is no advertisement of the Second City in the Southside. Nothing. And iO might just as well be called the "white ensemble theater" because goddamn everyone is white. We just got to advertise more. (Vaughn, personal conversation)

When we conceive of improv spaces as geographical, we may also consider the actual physical spaces designed for improv training, performance, and socialization:

> It's a question of pure infrastructure. You need stages; you need places to be, you need shows on a Friday night. If you're a poor black kid, you have no access to any of that shit. You might have access to a microphone to be funny in front of people, so you have stand up going for you. But you don't have theaters or anything like that. There is no financial structure as far as access to venues. You don't have that shit. Probably. I am speculating. (Schleelein, personal conversation)[12]

While this is a very white, rationalized point, I agree; if improv institutions and practitioners want to change their demographics, they will have to create this infrastructure.

The plan to open a Second City theater in Bronzeville in 2004 is a case in point demonstrating the challenges of infrastructure development. It also shows that spatial segregation is not only geographic. Allegedly "on a mission

---

[12] Derek Schleelein regularly performs at the Annoyance and is part of the Bastards of the Underground, both in their improv and their radio work. He wrote and acted in *The Thin Blue Line* (finalist in Comedy Central Short Pilot Contest) and can be seen in web series such as *My Bad Therapist* (produced by #Musenpet) and *The Chip Ganzler Show*. More film credits include the short films *Liminal Vortices* (SIGSALY Entertainment, assistant editor, script supervisor), *Moving-In-Law* (Studio Lumina Entertainment, production assistant) and *Fast Looker* (script supervisor). Our conversation took place in March 2015 in Chicago, together with Warren Phynix Johnson and Dacey Arashiba. Schleelein authorized the material used in this project via Facebook Messenger on 7 October 2019.

to better reflect the cultural diversity of Chicago" (Currie), the Second City planned for a theater and training center in the Southside that would "house five training center classrooms, a 150-seat theater, approximately four offices, and a bar facility for service of the theater" (Katz). Dionna Griffin-Irons, who was and is the director of the outreach program for Second City, described Second City's goal in this endeavor to "incorporate different points of view into their comedy" (Katz). The idea emerged from the same grounds as the creation and development of diversity and outreach programs in Chicago's improv institutions, starting with the Second City: namely the inability of their all-white casts to speak meaningfully on the subject of the LA Riots of 2002.

> "It was right around the time of the L.A. riots," Griffin explained. "Our actors were struggling to represent what happened in the riots. Then the idea kicked in that we have to have the African-Americans in order to represent what is going on. (qtd. in Katz)

However, Second City was not welcome in Bronzeville. In the anthology *The New Chicago: A Social and Cultural Analysis* (2006) Michael Bennett surveys criticism directed against the project, which was grounded in worries that "new investment in businesses and cultural enterprises may be intended not for their community but for the targets of an overheated real estate market." Moreover, the Second City's approach was perceived by some critics as a "racial and class-oriented encroachment into black cultural territory" (216).

The *Chicago Reporter*'s insightful cover story on the Bronzeville project – "New theater opening to mixed reviews" (Prince) – considers racialization in improv in a relatively nuanced way and includes a significant number of critical voices that address improv's whiteness. It is worth reading in full, but here are some excerpts:

> Some Bronzeville residents and community leaders say that Second City is taking advantage of their work to rebuild Bronzeville, and benefiting from federal funding intended for struggling communities, not successful businesses. And some members of Chicago's black arts community are wary. They point to the low numbers of African Americans on Second City's staff and stages, and to the track record of Chicago improvisational comedy being, as one black actress put it, "a white thing." [...]
>
> "It is a cultural slap in the face and a statement that we lack the capacity to develop that type of venue," said Harold Lucas, president of the Black Metropolis Convention and Tourism Council, a nonprofit Bronzeville-based organization providing entrepreneurial training to low-income residents. "I don't think that a European, North [Side], well-established comedy club should come into our community without broad-based support." [...]
>
> In March 2000, The Chicago Empowerment Zone and Enterprise Community Coordinating Council awarded Second City $850,118 for the Bronzeville theater. The Empowerment Zone (EZ) is a federal program that seeks to revitalize distressed neighborhoods.

> However, some community members oppose Second City's EZ award and its move to Bronzeville. They call Second City a "foreign" institution. "We don't need Second City to save us economically or culturally," said Lucas of Black Metropolis. Second City is "walking into a ready-made market," he said. "Groups have been working for 20 years to preserve the authentic heritage of the community."
> "This is a gross misuse of funds," added Nathan Thompson, a black author and activist. "Robert Townsend and Bernie Mac are the attractions that should be coming to Bronzeville, not Second City."

The white European mentality of conquer-to-save reverberates strongly here. Because of the ongoing racial dividing lines, whiteness is *foreign* to Bronzeville. The locals neither want nor need a white savior institution taking their federal money away from them. Everything about this sounds horrific, even though Second City announced some structural decisions in anticipation of such criticism:

> Leonard said that the Bronzeville theater [...] will be half-owned and fully managed by blacks. Its shows will feature integrated casts with a majority of the performers being black, said Dionna Griffin, the new theater's producer. She expects the new venue to draw racially mixed crowds. Half of the new theater's ownership structure will be comprised of nearly a dozen black producers, actors, directors and instructors who are alumni of Second City's training center, staff and stages, according to Second City officials. The other half will belong to Second City. (Prince)

However, for reasons difficult to adjudicate now, the "Black Second City" in Bronzeville "never materialized" (Lambert 200). Its cancellation speaks to the fact that a) questions of infrastructure can never be reduced to infrastructure alone, b) Second City did not do a particularly good job in communicating the project, and c) improv is viewed by Bronzeville activists as a business, an institutionalized entertainment rather than a politically idealized space for spontaneous egalitarian play. Such viewpoints are significant, as there is d) a political and individual affect in play that demands specific racial and racially cultural sensibility, which is difficult for an institution like Second City to generate or communicate.

Given the lack of infrastructure for improv, other problems logically ensue. One of those is the flow of information (or the lack thereof). People on the Southside of Chicago may simply not know about improv or encounter advertisements for shows or workshops. Perkins states that "When I went to college, I didn't even know what Second City was. My dorm went on a trip there and I was like, 'What is this thing that's been in my city in my entire life that I have never heard of?'" (personal conversation). For Patrick Rowland, it was the same:

> I grew up in the West side of Chicago, where it's all black community, and we never knew about improv. We know about stand-up comedy because of Eddie Murphy, Richard Prior, and all that, but when it comes to improv, people would always say Wayne Brady. That's all people would know about black people in improv. (personal conversation)

Information flow involves specific channels of communication and targeted (imagined) audiences. As Loreen Targos[13] suggests, a special effort must be made to advertise for an audience other than those informed by default (given their social sphere or general interest). It requires other avenues and methods:

> I went to go and see last year's Bob Curry fellowship thing. Me and my friends came to see it. It was less than five people there, so they canceled it. We thought that that was bullshit. What I think is that you need to do more outreach and marketing if you have a diversity show to get the same size audience. You need to push it harder and tailor it to minorities. Otherwise, they won't hear about it. If you do normal improv marketing, you will always only hit white people. (personal conversation)[14]

It would be misleading to identify these advertising barriers solely as geographical or due to a lack of awareness. Significantly, Vaughn considers the affective dimension that Black people may feel when wandering about in white spaces to see or to train in improv:

> It's so weird that people in Iowa can know Second City, but not people from the Southside, in the same city as Second City! I don't know how to bring awareness to them. A lot of the times, the parents are afraid to let their children out on the Northside or downtown. They're afraid of what can happen to their child. It's unfortunate that these children don't receive the same exposure to the arts.

Space structures libidinal investments like fear. The way one moves through the city reflects racialized segregation and racialized libidinal economies. Wandering around a white space is walking in continual physical threat, and so it is actively avoided. It may be far-fetched to suggest that the improv sphere as such threatens to kill a racialized person, but the universe in which it exists certainly does.

---

13 Loreen Targos performs with Stir Friday Night!, a nonprofit theater company founded in 1995. SFN is one of the nation's premier Asian-American comedy troupes and has been cooperating with The Second City and iO (see further the Stir Friday Night! website). Targos also performs stand up around Chicago and is a member of "PREACH," an improvised spoken word movement. She is a graduate of the Second City and Annoyance Theater programs and is a current student at iO Chicago. The conversation took place in March 2015 in Chicago. Targos authorized the material used in this project via email on 13 October 2019.
14 During my stay in Chicago, I specifically sought out shows with Black improvisers. Two shows were canceled because I was the only one in the audience; both had Black performers.

Black-racialized people may libidinally invest these white spaces with fear because they know that whiteness means violence. It may be an active, conscious, and sensible decision not to go there because such *safe spaces* are only safe for white people.

Other more pedestrian barriers exist as well, such as the distance from the Southside to the improv locations: getting to The Annoyance Theater, the iO, or The Second City from the Dan Ryan/95th (Chicago's southernmost expressway) takes almost an hour, plus the time to get to and from the metro stations. However, the ultimate barrier is of course financial:

> *Schleelein:* There is a very large barrier of entry as far as class is concerned. How many young black people are going to be like, "Yeah, I'm gonna take improv classes." You cannot afford that. There is a very distinct monetary barrier between the classes.
> *Arashiba:* To take the entire syllabus in Second City is going to cost you two thousand bucks.
> *Schleelein:* Two thousand, twenty-five hundred, something like that.
> *Arashiba:* Four hundred to five hundred dollars per class.
> *Schleelein:* Maybe it's worth it. It might be.
> *Arashiba:* That's also where you'd meet everybody else doing improv.
> *Schleelein:* But if you're a funny person in an underserved community who cannot drum up funds to take classes, you get on a stand-up stage. That's easy, that's right there. You're right there to do that shit.
> *Arashiba:* The barrier is lower. And all the black people I know who have come into improv have also done stand up first. (personal conversation)

Not only do people have to pay to get into improv (pay to play), but there is also no certainty that it will pay off in the long run.

> If you come from low-income families, a lot of times, you don't have the resources to make it in improv. Which is a lot of the reason why it's predominantly white. When I think about how much I paid on classes here before I got an internship – easily thousands and thousands of dollars. Not many people have that kind of money or that kind of time to make it all the way up here and take those kinds of classes, so it's a lot harder to get diverse talent. (Bullock)

Improv can thus be understood as a white economic luxury with little financial payoff, reinforcing one's perceived strength and sense of community and offering a privileged career path for those already privileged. It is a feel-good social luxury often impossible to attain in the white sphere for anybody racialized as Black. In other words, improv is a path to further achievement in the entertainment industry for those privileged by whiteness.

Staging segregation

The factors identified above speak to the racialization of space as discussed by Charles Mills in *The Racial Contract*. He describes the normative racialization of space on three dimensions: "the *macro*level (entire countries and continents), the *local* level (city neighborhoods), and ultimately even the *micro*level of the body (the contaminated and carnal halo of the non-white body)" (43–44). While the normative characteristics ascribed to these spaces have been discussed at length, it is worth pointing out that these dimensions are not only linked but, in political epistemology, create and reenact each other:

> The norming of space is partially done in terms of the *racing* of space, the depiction of space as dominated by individuals (whether persons or subpersons) of a certain race. At the same time, the norming of the individual is partially achieved by *spacing* it, that is, representing it as imprinted with the characteristics of a certain kind of space. So this is a mutually supporting characterization that, for subpersons, becomes a circular indictment: "You are what you are in part because you originate from a certain kind of space, and that space has those properties in part because it is inhabited by creatures like yourself." (41–42)

When we look at what is quickly framed as segregation, we must keep in mind that such segregation results from a racialization of space that is manifest in other dimensions, such as the individual body. We must not give in to the idea that Chicago, one of the most segregated cities in the US, has a problem that can be solved by establishing or improving an "improv infrastructure." Because this is so, Second City Bronzeville could not have happened; the psychic, political, and moral structure that gave rise to the project mirrors "European cognition," as Mills would say, which assumes that

> in certain spaces real knowledge (knowledge of science, universals) is not possible. Significant cultural achievement, intellectual progress, is thus denied to those spaces, which are deemed (failing European intervention) to be permanently locked into a cognitive state of superstition and ignorance. (44)

Second City's attempt to build a theater in Bronzeville performed such a "European intervention" with very "European" moral, financial, political, and missionary intentions, and was consequently rejected by the people inhabiting that space. However, the characteristics of the racialized space – the impossibility of real, i.e. European, knowledge – are also applied to the individual racialized body. Vaughn considers the racialization of space *within* the improv community:

> We can start a program where it's like "all people of color," and we're gonna have a scholarship for them and put up a sign and have an intensive for them and showcase the work. But all you're doing is segregating them as well. So, you can't win in that. And then they get mad because essentially it is kind of affirmative action. Getting us scholarships so our voices can be heard, because obviously, those white improvisers are not trying to get us out there. They don't really believe in what we have to say. They think it's gonna be all chicken jokes and talk about things that don't matter to them. But we have voices. There are people here who have never been to the Southside and do not know that it's not that bad. You can get stabbed at the Northside. Wrigley is fucking crazy. Everyone is labeling the middle easterners as terrorists. It's the white men who are killing people. (Vaughn, personal conversation)

The concept of Black "voices" comes up frequently, and with it the observation that non-white voices are not deemed relevant for white improv comedy:

> With Stir Friday Night!, we definitely had Asian people come more to shows. I do suppose that white people feel like "This isn't speaking to me." Same with other minorities. Black theater will attract mostly black audiences. White people do not feel like it applies to them. All of us [the Stir Friday Night ensemble] play in other groups as well. And in all those places, our stories aren't as celebrated or centralized like in ours. In our group, we can tell our stories, and there is a synergy. Within our group, we have similar stories, or even if we don't, we better understand the need for you to tell your story more than if you were in some white group. It's just different. (Targos, personal conversation)

Even within an improv team, then, the individual body is racialized in that it is *intuitively* restricted to characters assigned to racialized spaces that cannot hold *real* knowledge, for example by giving them ghetto names:

> So often you would watch a show with an improviser of color, and you would find that person in an awkward situation, an awkward scene in which you're like "I wish they wouldn't –" Like you feel embarrassed for them. A lot of times, it's not malicious, but if you're not sensitive to it, you say things that are annoying to people of color onstage. Like when you give them "ghetto names" subconsciously. It's always rewarding to me to see people of color in an awkward situation with a racist cast member – and how they can flip it and turn it into a beautiful scene. (Bullock, personal conversation)

> I came into a scene, and I did a British accent. And the first thing my scene partner says: "There are no black people in London!" Which a) there are tons of, and b) why would you say that? Every time I come out in a scene, they automatically either label me as "Tyrone" or another "black person name." Or they would unmake me: "This guy will be the black guy." They probably haven't dealt with a lot of people of color, so all they know is what they see on the news, what they see on TV, and that's their only way to filter what they're gonna say in the scene. And it's sad that that still happens. (Rowland, personal conversation)

Consider the verb "unmake," which Rowland uses to describe the dynamics of such on-stage racialized labeling. This speaks to the way in which he is being confronted not with "discrimination" or a "stereotype," but ultimately with the repeal of his ability to play from "the top of his intelligence."[15] His fellow white improvisers will, consciously or subconsciously, assume his mind can never harbor the *real knowledge* of white cognizers. Black-racialized improvisers are assumed incapable of telling meaningful stories, or even universal ones, but are thought to be bound to meanings of Blackness:

> When the groups or shows aren't segregated, it becomes such an eggshell type of situation. With sketch, you have a little bit more control, but with improv, everyone is walking on eggshells. There will be scenes where just because one person in the scene is black, it changes the tone of the entire scene. I definitely know there are scenes that I have done, in which I got a laugh – or not a laugh – because I was black. I remember a scene in a freeze tag jam. I tagged out somebody who was on the ground, and I was on my knees. The first line from the other two people, who were my friends, they didn't mean to do this, but they said: "Oh, he's ours now." And I thought I was being an animal. I wasn't even thinking a racist thing; I was being an animal that they found. And then immediately, the crowd goes "Ooooooooooh." And then we realized what they saw. They thought it was a slave scene. It just comes up in weird situations like that. (Boyd, personal conversation.)

Unlike Boyd, when asked whether race is always there once a Black improviser enters a scene with only white people, Johnson answers in the negative.

> *Schleelein:* Warren, you say no?
> *Johnson:* If a motherfucker makes you a slave in a scene. What you do as a slave in the scene is to make yourself a real human being. "Gerald, I told you to take that trash out yesterday." "Sir, I am going to take that trash out. But I am just so conflicted by how your wife can smoke a cigarette – and she's pregnant."
> *Arashiba:* Don't make slavery the thing. The thing is still the relationship between the two people.
> *Johnson*: Exactly.
> *Schleelein:* That's how you get through it and expose how stupid it is, too. I don't have any way to relate to that premise except for the fact for the idea that you want to destroy that idea as soon as it happens. And I do think that's true: If you have a white person and a black person on stage, immediately it's gonna be some consideration of race – on any level. You're gonna think about it when those two people get up there. They can be married and

---

15 This is a common phrase in improv teaching attributed to Del Close. In a blog entry, long-time performer and instructor Jimmy Carrane states that for him, it means to make a "choice that comes from honesty, that reflects life." I am making this point here because the notions of *honesty* and *life* are common tropes in improv poetics. However, performing truth and honesty is a hard task for the improviser who is put into characters with their (always already putative) humanity *unmade*.

immediately smash all the preconceptions you might have about black people, especially. But I think it's undeniable that when you see it, that it's there. I really would love to see a scene where that is not even acknowledged, but everybody always does it.

Johnson: Not always, that's not my thing.

Schleelein: I'm not saying you personally.

Johnson: Every scene I go in, I am a white guy. Every scene I step into, I am a white guy.

Schleelein: Unless someone tells you you're a black guy, you are white.

Johnson: My usual thing is a white guy. But yeah, unless I'm defined –

Schleelein: I love it when the shit goes the other way. We did an improv show where it's just you, me, and somebody else, and you were like, "Yeah, we're all here, just a bunch of white guys."

Johnson: I am always the white guy in my scenes unless I am a) defined by somebody else, or I define it as myself. Always assume I'm a white guy, always. In all my scenes, I am always white because it's a neutral guy.

Mostly, when Black-racialized improvisers share anecdotes and perspectives on the subject, they make the point that the racial behavior of their white scene partners is "not malicious" (Bullock, personal conversation). However, these questions are not moral, because the Racial Contract creates such morality in the first place (per Mills). We need to remain in the sphere of structural analysis. It would not be helpful to consider what is or is not *allowed* on stage – by whom, for whom, on what grounds? – or who is or is not racist. Let us keep in mind that I am addressing a fundamental anti-Blackness from which nobody can escape, least of all white improvisers.

We must understand that Blackness matters, means, and signifies on the stage, and that we can frame this in Mills's concepts of the multidimensional racialization of space that structures the world on the macro-, local, and micro-levels. Accordingly, dehumanizing ascriptions of Blackness onto all of these spatial dimensions follow the same rules and dynamics. That is why we must understand how integration and segregation are not mutually exclusive but reciprocally affirmative. When we speak about segregation in improv, we must understand it in these various dimensions, including segregation within improv, within specific institutions, within groups, and between scene partners. An all-Black house team can exist in segregation. A Black improviser in a group can still feel segregated, which they may or may not have the interest or ability to "turn into a superpower" (Boyd, personal conversation). The issue of racialized, segregated spaces persists even when Blackness is absent from the white geographical and intellectual space of improv, and when it is present, improv remains segregated from within. The failure to recognize the multidimensional racialization of spaces is why the politics, rhetorics, and practices of diversity in improv and elsewhere are doomed.

Efforts at improvement

Second City's outreach programs, motivated by the LA Riots in 1992, were the first in the improv world. Other companies have followed. As of 2020, Second City has broad offerings that range from scholarships and festivals to the central Bob Curry fellowship, a "ten-week master program that offers up to 16 fellowships to qualifying actors and improvisers from diverse, multicultural backgrounds to train and study at The Second City." Co-funded by NBC, the Bob Curry fellowship is tuition-free and "awarded to exceptional talent selected by audition and adjudicated by Second City directors and producers," as presented on Second City's website ("Diversity"). Additionally, Second City offers up the decidedly diverse ensemble "Urban Twist," formerly known as "BrownCo":

> It's basically an accelerated process of a lot of the work they have at Second City. It teaches you how to put together an archive show, how to write it. You get paid a $400 stipend, and you meet maybe two or three times a week, work on material and then put together a show that producers get to see. That's a really nice opportunity because it's specifically for people of color or diverse talent, and you don't have to spend years and years and years which you may not have the time to invest loitering around at Second City. (Bullock, personal conversations)

However, more than two decades later, Patrick Rowland observes the same problem in the context of ongoing anti-Black police murder:

> We're representing people's lives. We're representing their reality. You can't do that if you don't have a diverse enough cast. So you can't speak on stuff that's happened in Ferguson. Stuff that happened to all these black teens getting shot. It's hard to do that from a white perspective. It's a lot easier to bring in people who have that voice, who've grown up in that neighborhood, who have something to say about that and can put their own spin and their own voice to it. (personal conversation)

Looking back on many decades of professional improvisation in Chicago, TJ Jagodowski states that racial discrimination in improv is less of an issue than gender discrimination. He also observes a change in the casting processes, politics, and sensibilities over the years:

> I don't recall racial discrimination in my personal improvisational history. I do recall some quiet gender discrimination. When outside teams were put together or casts aligned, the presence of a female performer was often an afterthought or an add on because "we need to have a woman." Also, it was a very different time for the LGBTQ community when I started playing. It was the generation of "Don't Ask, Don't Tell." That basic tenet permeated most of American life, so there was inherent discrimination against gays being able to be out and be proud. It was something that wasn't talked about very much. I don't recall any discrimination that I've come across in workshops or on stage. Di-

versity plays a role for certain now in casting. It would be unusual to assemble a cast of any size that is all white, all-male, or all straight. In many ways, because it limits the scope of ideas that can be presented with proper representation or with personal responsibility. (personal conversation)

Second City is not alone in its "diversity efforts." Other theaters like UCB and iO offer forums, meet-ups, panels, diversity jams, and make other efforts of outreach. (See their respective websites for more.) Rarely, however, do the advertisements of diversity and inclusion programs consider or formulate the lack that motivated them in the first place. The argument is rather a transumption of the "cultish" origins of improv. Consider the Second City's value statement:

The Second City believes that everyone, regardless of race, gender, or sexual orientation should experience the gift of improvisation and the many life skills it helps develop in a safe, supportive way and is actively working to champion diversity, equity, and inclusion at all levels of our institution and associated programming. Our goal is to open doors, create a bridge, and make this work accessible and inclusive. We firmly believe that unique experiences and points of view are at the core of the art form our artists create and reflective of the audiences, students, and clients we serve. ("Diversity")

While improv, here institutionalized as Second City, harbors the problem and has always done so, it nonetheless presents itself as the solution. Being so fundamentally white is slowly understood to be an ethical dilemma; this is a lack on the part of improv, not on the part of those absent, who are implicitly blamed for their absence or even ascribed a lack of "life skills" or "bridges." But bridges to where? As if that twist in logic wasn't problematic enough, the talk of diversity and inclusion throws together heterogeneous groups, which it subsequently creates as *homogeneously diverse* in performative praxis. The DiOversity Scholarship at iO is "offered to POC, LGBTQIA, and disabled individuals in order to complete the iO improv program." It is presented as "one of many ways that iO is trying to ensure that more marginalized voices and performers from communities underrepresented at our theater, and in society at large, have a chance to get on iO's stage and put great comedy out into the world" ("DiOversity"). POC and disabled individuals do not necessarily share anything except not being in the position to form their own world.

Of course, this register of diversity is not specific to improv, and I examine it further below. Let us look briefly at how the rhetorics and practices of diversity are perceived within the community. According to Chinyere, at UCB New York, the debate about and programs for diversity fail to provide a grammar for visible change (not to speak of that which is "untalkable," to use Boyd's term):

Nearly six years into my stay at UCB and it is very much unclear what the purpose of the Diversity program is at UCB. Is it to remain compliant? To give the illusion that they care about diversity? I have taken several workshops and 17 classes at UCB since 2011, a few of them were paid for by this scholarship. So it can definitely help finance your journey through UCB – but that's only if you choose to stick around, re-apply and get selected.

You see, a lot of people who accept these diversity scholarships do not in fact stay, so much so that the Diversity program had to incentivize students with a second free class if they took the first within six months or year. Why, you ask? Because a lot of POC show up the first week of class – don't see a lot of folks like them and drop out. For context, I was the only person of color in my first four classes at UCB.

Why doesn't it "work?" The rhetorics of diversity represent the only register applicable and the only political strategy available. This is not a moral question or a political critique but rather speaks to the inevitability of racialization in improv discourse and practice – including the fundamental shortcomings and even more problematic assumptions they entail. What we need to analyze lies far deeper than the intelligible arguments considered here, so far from the thinkable that language and any strategies to do something about it keep failing. According to Vaughn, the concept of diversity is not a solution, but part of the problem:

> In Chicago, there is very big segregation going on. And in the improv community, it's still that way. Even though we sometimes say "Hey, we need diversity" or "Hey, we need a black person or an Asian person" or whatever. But there's so not a voice. If you have to say that we have diversity here, obviously you don't. If you need to put a sign up saying "Diversity – YAY!", you don't have diversity. And why does it have to be that way where it is so segregated, and you need a white man, or you need a white woman, but you don't need more than just one black person's voice. Each of us have different voices. Each minority does. You can't have one minority represent all; that's unfair when you have three or four white guys up there. (personal conversation)

Whereas Bullock thinks of the Bob Curry fellowships as "a really nice opportunity because it's specifically for people of color or diverse talent, and you don't have to spend years and years and years which you may not have the time to invest loitering around at Second City" (personal conversation), Boyd makes a strong point about its ambivalence, finding segregational diversity at once "a good thing" and "aggravating":

> Everything is segregated. Sometimes it's a good thing. Most of the time, it's not. Just the fact that there is still "Gay Night" and "Black Night" at a lot of theaters. And I get it because sometimes you have to market shows like that in order to find that demographic of people who would be comfortable watching that type of comedy. But the fact that Second City does have a diversity and outreach section that's really thriving and finding a lot of good minority comedians in Chicago is good because it is an avenue that we probably wouldn't have had in the way that some improvisers come to move here. I definitely think that there are

people who just don't know what steps to take to get them from A to B. I have a lot of friends who have day jobs, kids, they are trying to go to school, they would never even have heard of the opportunity to take classes at Second City in order to do improv. They wouldn't have known that was an option for life, to pursue that as a career or hobby. So definitely that sector of Second City is doing great work. That's great. But it's also aggravating. (Boyd, personal conversation)

In contrast, Loreen Targos recognizes what for others may be a structural problem, but values its pragmatic effect over its negative repercussions:

I do see how people say it segregates the community. But it's segregated anyway. Before these groups were there, it was even more segregated. At face value, it seems like "Oh, you're separating yourself!" but I think people have a right to talk about their experiences and be in a space where they feel safe to express a particular part of their identity. A lot of the time, the majority does not make improv a safe space. (Targos, personal conversation)

Whatever these heterogeneous positions on diversity, white discourse ultimately has the prerogative over how it is represented, understood, coded, and judged. Consider this anecdote from Targos:

There was a group that I was in, and we were surprisingly diverse, accidentally. We thought it would be funny to call ourselves "Affirmative Action" because we were so diverse. And our coach, who was white, was like, "No, you shouldn't do that. People are gonna think that you got picked for this because of affirmative action, and that has a negative connotation." For us, it didn't have a negative connotation. He thought that for our audience, which was mostly gonna be white, it'll have a negative connotation. So we thought of something else. Something generic. (personal conversation)

Failing rhetorics of diversity

This project is flawed in that it is bound to a grammar it seeks to critique. In looking at diversity policies – potential ways to *get more minorities on stage* or to *reduce discrimination* – I would be theorizing, presenting, and arguing in a register that doesn't uphold these concepts in the first place. In a conversation with Frank Wilderson, Jaye Austin Williams considers the limitations of diversity, which cannot address the anti-Black structure of Western modernity:

[I]t's not enough to focus on "diversity" or to talk about racial hatred and discrimination in their broader contexts. Rather, it's about how anti-blackness constitutes a structured antagonism that manifests overtly and insidiously, and in turn, impacts everything else – not least, the psychic and physical lives of black people all over the world. ("Staging" 2)

Diversity as rhetoric, politics, and practice is mobilized to assuage the ethical dilemma of the white subject. Wilderson writes in *Red, White & Black*:

> At the beginning of the twenty-first century, the irreconcilable demands embodied in the "Savage" and the Slave are being smashed by the two stone-crushers of sheer force and liberal Humanist discourses such as "access to institutionality," "meritocracy," "multiculturalism," and "diversity." (30)

The concept of diversity comes from a place of default violent whiteness while presenting itself (and being imagined) as an alternative, as progress. It appeases mainly white liberals or progressives who would be threatened by the consequences of recognizing the anti-Black violence that makes up (their/our) white subjectivity – regardless of how empathetic they think themselves when it comes to a generalized discrimination. As Wilderson observes:

> A lot of fear rears its ugly head among our so-called "allies" when we move toward that something deeper you're talking about; something deeper than inequality and injustice, as you point out. I think of it as a fear Left-leaning artists have of what will happen if we stage a black encounter with violence. They don't want to witness our singular relation to structural violence – start in with that and they will burn you alive, as a witch at the altar of the universal. (Williams and Wilderson 28)

Liberal and progressive concepts like multiracialism or diversity rely on the belief in minorities and ethnicity, whose function Karen Fields and Barbara Fields describe as a feature of what they term "racecraft." They differentiate "racecraft" from racism and race: "the term *race* stands for the conception or the doctrine that nature produces humankind in distinct groups, each defined by inborn traits that its members share and that differentiate them from the members of other distinct groups of the same kind but of unequal rank." For Fields and Fields, racism denotes "the theory and the practice of applying a social, civic, or legal double standard based on ancestry." It does not speak to an "emotion or a state of mind, such as intolerance, bigotry, hatred, or malevolence" but is "an action and a rationale for action, or both at once." It takes "race" for granted, but the terms are not the same. Whereas "racism" describes "something an aggressor *does*," race is believed to refer to "something the target *is*." (16–17)[16]

---

[16] In the ensuing chapter, I will go a step further – denoting that "race" is also a concept that transumes "Blackness" to avoid confrontation with the anti-Black structure in which modern subjectivity is born. Talk about "race" again becomes secondary, because I am not at all interested in – or even suggest anything could be said about – "what a target is," and the notion of "unequal rank" will be left aside, because Blackness ultimately does not rank at all. Anti-Blackness, though raced like other "ethnicities," has a more fundamental function. Racism as an "action and a rationale for it" will not be part of the discussion, because the idea of a rationale will no longer be relevant. However, at this point in the argument, I firmly believe that the reader new to such a line of reasoning will find this transition helpful. Additionally, the concept

What they call "racecraft" does "not refer to any groups or to ideas about group's traits" but denotes "a mental terrain and [...] pervasive belief." It can be objectively observed and has "topographical features that Americans regularly navigate," originating not in nature but in "human action and imagination." Racecraft is not racism; it is "fingerprint evidence that *racism* has been on the scene" (18–19). Terms like ethnicity or minority function discursively as substitutes for race, and are complicit in veiling the racist dimensions of the discursive grammar that makes them semiotically and affectively intelligible in the first place. Like a thick gauze curtain, in their lack of precision and in their powerfully immediate affective understanding, these more malleable terms provide a way for white liberal progressives to talk about race without having to acknowledge its racial foundations.

Fields and Fields point out the logical absurdities that follow from muddling the individualized and the quantitative dimensions contained in the concept of "minorities," which veils structural racism qua its function in and as racecraft:

> "Minority" ranks alongside "the color of their skin" as a verbal prop for the mental trick that turns racism into race. The word slips its literal meaning as well as its core definition, which is quantitative. Vice President Spiro Agnew once demonstrated the trick unconsciously. Responding to a question about American policy toward the white supremacist regime in what was then Rhodesia, he said it was no business of the United States how other countries dealt with their "minorities," by which he meant the country's black majority. The quantitative meaning slips again in the paradoxical formula "majority minority," referring to the projected numerical predominance of non-white persons in the United States in the not-so-distant future. If the logic were harmless, it would be hilarious. (28)

The flawed notions of minorities or ethnicity are nonetheless part of the foundation that creates the progressive concept of diversity in the first place. They are complicit in the maintenance of the racial system that brings them about and by definition can never work against it. They cannot be mobilized to alleviate Black suffering, because liberal thought, as it perceives itself as progressive, can never work toward its own ruination. In structural terms, diversity as a concept, a practice, and a political prerogative is an essential part of racecraft and functions to make racism, and anti-Blackness specifically, invisible to those who believe in it. The register of diversity involves a relationality between ethnicities and cannot provide a grammar for the kind of change it explicitly claims to foster. If (left) liberal progressive discourse could see the world without ignoring the anti-

---

(and modality) of the idea of "racecraft" as a mental terrain is helpful. I think this helps many readers to follow through to understand how "diversity" as a plurality of ethnic (or other) identities is in itself a racial concept, and, as such, useless to attack racism.

Black violence that structures it, it would have to annihilate itself. As a concept and a term, diversity says as much by involving the anti-Black vector, pointing away from a default position inhabited by those *not-diverse* and throwing together the least-connected positions and non-positions provided by social life and death. When the iO offers scholarships to "POC, LGBTQIA, and disabled individuals," it is not doing much against racism in and Black absence from improv. Rather, it contributes to it by maintaining a social stability that is a "state of emergency" for Black-racialized people (Wilderson, *Red* 7).

Improv and the politics of hope

In his essay "Black Nihilism and the Politics of Hope" (2015), Calvin Warren argues that "[e]very emancipatory strategy that attempted to rescue blackness from anti-blackness inevitably reconstituted and reconfigured the anti-blackness it tried to eliminate." The rhetorics of diversity are located in the larger discursive vector he terms the "politics of hope" (239). Hope exists only for those who are understood to participate in the Political – a sphere which is by definition foreclosed to those racialized as Black:

> The logic of the Political – linear temporality, biopolitical futurity, perfection, betterment, and redress – sustains black suffering. Progress and perfection are worked through the pained black body and any recourse to the Political and its discourse of hope will ultimately reproduce the very metaphysical structures of violence that pulverize black being. (218)

Such a frame helps to articulate the philosophical axioms activated here, showing that diversity politics are not in fact as paradoxical as they might appear because Warren enables us to theorize a "hope" that "pulverizes black being." To begin with, we must differentiate with Warren "between 'hope' (the spiritual concept) and the 'politics of hope' (political hope)" (218); while the former may "always escape confinement within scientific discourse" by convention and through discursive compulsion, it still attains "intelligibility and efficacy within and through the Political." With Grant Farred, Warren argues that "political participation is motivated by self-interested expectancy; this political calculus assumes that political participation, particularly voting is an investment with an assurance of a return or political dividend" (219). Such a return or dividend becomes a (Lacanian) "impossible object" for Black non-citizens (221), which makes Black political participation irrational "given the historicity of voting as an ineffective practice in gaining tangible 'objects' for achieving redress, equality, and political subjectivity" (220):

> This idea of achieving the impossible allows one to disregard the historicity of anti-blackness and its continued legacy and conceive of political engagement as bringing one incrementally closer to that which does not exist – one's impossible object. (221)

Warren suggests that recognizing this impossibility within the field, grammar, and engagement of the Political is hidden behind the logic of problem and solution that structures it. For every problem, the Political offers up a solution. When we look at the Politics of improv, for example, it is assumed that more exposure and representation, woke teachers, more diversity, or simply better infrastructure would solve its *problem* of anti-Blackness (as manifest as Black absence, segregation, or discrimination). It is rare that anybody would say "This cannot be solved."

Warren's concepts of Black nihilism and political hope theorize just that impossibility. They provide a clear view of the fact that a system is not capable of (or interested in) "solving a problem" that is constitutive of that very system. Political hope serves to veil that incapacity by continually projecting the solution into the future. Warren argues that the "*idea* of linear proximity – we can call this 'progress,' 'betterment,' or 'more perfect'" is the axiological ground for what he terms the "'trick' of time:"

> Because the temporality of hope is a time "not-yet-realized," a future tense unmoored from present-tense justifications and pragmatist evidence, the politics of hope cleverly shields its 'solutions' from critiques of impossibility or repetition. (222)

This is the ground for every "good" action that intends to combat the problem, because *taking action* is related to *problem-solving*, as inaction will not change anything. However, the solution is dialectically dependent on the problem: "The politics of hope, then, depends on the incessant (re)production and proliferation of problems to justify its existence. Solutions cannot really exist within the politics of hope, just the illusion of a different order in a future tense" (223). Further:

> To "do something" means that this doing must translate into recognizable political activity; "something" is a stand-in for "politics" – one must "do politics" to address any problem. A refusal to "do politics" is equivalent to "doing nothing." [...] To refuse to "do politics" and to reject the fantastical object of politics is the only "hope" for blackness in an anti-black world. (223)

The improv problem under investigation here can be *solved* in the sense that the solving as praxis is a means without an end. The "doing the solving" is the solution in itself, because in *doing*, improv can claim to be doing *something*. The entire diversity project and its political engagement must be read in these terms. Such empty *doing* does, of course, solve the problem that white people have with Black absence. Structurally speaking, it does nothing if not "aggravate" (Boyd) the anti-Black structures that Black improvisers regularly experience in their various improv-specific manifestations. It does not matter whether all this doing actually leads to something. Speaking with Warren, the politics of hope are even more efficient if it does not. From Black *non*-positionalities *not* solving it, staying away from this doing, would be the only "'hope' for blackness" (223). With Warren, we cannot but acknowledge that "political hope is bound up with metaphysical violence, and this violence masquerades as a 'solution' to the problem of anti-blackness." We must understand the axioms and concepts of "temporal linearity, perfection, betterment, struggle, work, and utopian futurity" as "instruments of the Political that will never obviate black suffering or anti-black violence" but as tools that "only serve to reproduce the conditions that render existence unbearable for blacks." This problem cannot be *solved* in the realm of the Political. It cannot be solved without *ending the world as we know it*, as Wilderson frequently puts it. Asking "What can be done?" is an exacerbation. Warren posits "political apostasy as the spiritual practice of denouncing metaphysical violence, black suffering, and the idol of anti-blackness. The act of renouncing will not change the political structures or offer a political program; instead, it is the act of retrieving the spiritual concept of hope from the captivity of the Political" (243).

The discursive environment and climate of improv is passive aggressive towards such fundamental criticism. Based on notions of improvement, ignorant improvisers will always prefer to fashion themselves as *woke* rather than acknowledge that their wokeness cannot be trusted. Making such statements, however, always hints at improv comedy "becoming better" in itself, or even playing a role in instigating a debate aimed at the betterment of society:

> It feels like over the past six years, people are becoming better about it. I don't know why that is. Maybe because people started shutting that stuff down. But it still shows up. Nobody wants to control what happens off stage. Or be the determiner of what is correct and what's not. But at the same time, people are starting to learn that they have limits. (Arashiba, personal conversation)

> I feel like comedy is one of those things that allows people to talk about those risqué topics in a fun way. Or at least when there is a joke on top, it's easier to hear about racism, sexism, classism or homophobia, all that kind of stuff. (Boyd, personal conversation)

Patrick Rowland articulates that ambivalence in the language and grammar of negotiation and progress, but is not particularly optimistic because he understands its ultimate aim as Warren's "trick of time":

> I was really happy when I heard about your project because I think diversity in Improv is still barely existent. It's crazy. When I started taking classes, I didn't see any black people or any people of color on the stages, and there wasn't really anybody of color in my classes. It's kind of getting better; I can at least count on two hands black people on teams here, but it's still slow progress.
>
> I think it's just exposing more people to us, to people of color, having more of us doing shows and teaching. Like right now, I think Nnamdi is the only one of color here at iO teaching. And hopefully, I'm gonna start teaching soon. I've gone through the program at Second City and at iO and some at Annoyance, and I have never had a teacher of color teach me. It's just like if you have people already like teaching, able to nip that stuff in the bud when people are learning improv. I think it would start making things better. In an ideal world... I don't know if it'll ever happen. Sadly. I don't think it'll happen. (Rowland, personal conversation)

Through this sketchy report of anti-Blackness in improv, I now arrive at a simple point. It is surely a long stretch to consider every Black-racialized person who *does not do improv* a Black nihilist in this sense. But it does help us white people to understand why the improv we cherish is so white. And it forbids us to seek explanations of Black absence in the spaces, meanings, and individuals that inhabit the sphere of Blackness, unless we appreciate it as an active and conscious choice, and the only rational one to be made: a kind of improv apostasy in advance (or in retrospect, as some examples above demonstrate).

In improvisation itself?

When presenting the thesis of this project, I have repeatedly come up against one question or defensive argument: "Yes, that may all be true. But do you really think improv is worse than other aspects of life in that regard?" Most white improvisers are annoyed that I have become a *race traitor* (in their eyes), and do everything they can to defend *their* improv against the positions I have articulated. Such defenses miss the point completely. I am not interested in the morality of improvisers, in the potential of improv as an art form, or in strategies for its *improvement* in the *solution* of the *diversity paradox*. I am not attacking anybody individually any more than I am attacking myself, and I made a living on improv for several years and still perform sometimes. In addition, I cannot be sure that all (or any) of the arguments I am making would resonate with the Black-racialized improvisers and performers of color to whom I have spoken. Yet it is not my aim to find one coherent theorization of improv, to level the plurality, to eradicate ambivalences that the anti-Black foundation of modernity generates for people living in the real world. When asked about improv's specificities in this regard, Loreen Targos rejects the idea of improv being somehow special, and

David Pasquesi shares the common white belief that improv, or improv training, may even hold some kind of a remedy to *better* the world, ridding it of "prejudice:"

> I don't think people in improv are necessarily more or less enlightened than the general population. I think, if they are ignorant, it's a lot easier for them to demonstrate that ignorance because they're on stage being an idiot. (Targos, personal conversation)

> I am aware of discrimination... that there is discrimination everywhere. I feel that I personally do not discriminate based on things like race, gender, sexual identity, religious beliefs, nationality. But I can't be sure that I don't. However, I definitely discriminate based on lack of ability or talent. There definitely are prejudices in the world and in the world of improvisation. But one nice thing about our little world is that things like prejudice are antithetical to being a good improviser. Inclusion, acceptance, and embracing differences are helpful in improvisation. So I think the problem with discrimination in improvisation lies more in the administration and business hierarchy and not in improvisation itself. (Pasquesi, email follow-up to personal conversation, 20 Nov. 2019)[17]

In my opinion, the task of the scholar is to maintain their legitimacy by respecting any position that does not conform to their theory. Freedom of speech, so to speak, does not entail freedom from disagreement, and debates about race are rarely without disagreement.

I suggest that improv, imagined as an aesthetic modality and practice, may not be immoral per se as long as we conceive of it purely as that. Yet such a perception is insignificant because a) there is no such thing as a pure aesthetics that is never realized, and b) morality and ethics are concepts and beliefs located in the assumptive logic of the Racial Contract and thus wedded to the Political. Improv is realized where people perform it – on public stages, in classrooms, wherever. And if we talk about it, we must concern ourselves with what actually happens.

> I am pretty sure everybody has a story of that where they brought a friend and then felt embarrassed by what was onstage. It happens so much. It happens to everyone. It's almost a common thing. (Bullock, personal conversation)

> At a Second City show [...] the Nommo Gathering Black Writers Collective, the only ensemble of black writers to go through Second City's training program, watched white actors per-

---

17 A note on "talent": for Wilderson, meritocracies are "stone-crushers of sheer force and liberal Humanist discourses" (*Red* 30). In the discussion of anti-Blackness, the idea of talent is a powerful derailment that blames the absent in an essentialist way. Aaron Freeman, the first African American performer at Second City, always knew that the "definitive question in a scene wasn't my talent, but what was appropriate to my skin color. It was a distinction that was made. That hurt" (qtd. in Prince).

form a parody of the drug culture, said Stephanie Shakur, the group's executive director. "One woman was holding up an infant, giving it a bottle. Another walked up, with the stereotypical black cadence, and said, "Girl, what's wrong wit' you?" The other woman pointed at the baby and said, "Girl, she pregnant!'" said Shakur. The mostly white audience hollered with laughter, she said, but a dozen or so blacks from the Nommo Gathering sat in stunned silence. (Prince)

In an advanced study class, my teammates once made the choice for me that I was stealing from a store. All I did was walk into a scene, alarms sounded, and I was arrested. What does that say when the only person of color in a scene is immediately arrested? (Chinyere)

It's so weird talking about it because we see it all the time. And I always have to remember that not everybody watches improv the way I watch it. And I watch improv as a person of color – "land mine over here, land mine over there." You're tiptoeing. And the more you do it, the less authentic you are as an improviser. It's stifling. (Bullock, personal conversation)

When I'm on stage with a white actor, their first idea is "What's up, homeboy?" or – "You're in jail, and I'm the police officer," or they call me "Malcolm X." They already typecast me as a black person. If I have my hair in an Afro, that makes it even worse. (Ronnel Taylor qtd. in Prince)

I was at Second City auditions for level 3. A friend of mine asked me, "What advice can you give me?" I was like, "First of all: be human! Don't be crazy all the time. And then: What is special about you? Showcase you." And she was really nervous because she was the only person of color within that group. She was like: "I don't know how to showcase myself and not make myself feel isolated." And I was like, "Oh, that's a problem that we all have at some point. But the more you do it, the more you figure out that you can use all of you as a human on stage. You don't go on stage as a blank slate. You have all your experiences as a person. So if you come upon something, just respond like you respond in real life." (Perkins, personal conversation)

All these anecdotes point vaguely to a field of investigation covered by phenomena that cannot be solved by the establishment, by improving infrastructure, or by increasing exposure for Black performers. They hint at something more fundamental than prejudice, something that is not specific to improv but addresses our understanding and mobilization of Humanism.

Here are my answers as to why improv's specific modal, cultural, historical, and libidinal configurations can and should be analyzed in their specificities concerning anti-Blackness. First, improv discourse (and its self-description) is so heavily built on a politics of hope, and is such a left-liberal progressive idea – and has been from its beginnings – that there is a specific need to address the anti-Black violence that undergirds it. Improv can be understood as a prime project of the politics of hope. The notion of betterment through improv can be found on the microlevel of improving individual personalities with effects on their personal surroundings, because "improv forces you to listen [and] teaches empathy and emotional intelligence" (Naameh; see also Streu and Wilson).

There are also larger projects like Improv for Humanity, which believes in applied improvisation "to address the vast array of challenges facing us – from refugee crises and climate change, to disaster preparedness and working with vulnerable populations" ("Who"). With such powerful hope in the discourse, a thick veil disguises the violence that is the modern foundation of structural anti-Blackness. Second, as considered above, improv has recently discovered active *problem-solving* in the politics of hope. The debates that surround it are gaining some sort of (temporary?) momentum, and we can expect that anti-Blackness will soon find ways to recast itself. Third, and most importantly, several fundamental axioms and dynamics that constitute (and define) improv's poetics affect the reenactment of white supremacist anti-Blackness. Improv lore (both practical and academic) holds that improv can overcome old forms of knowledge by referring to a spontaneous reality somehow generated via access to intuition, the practice of play, and the mobilization of humor. I suggest that all these fields provide a specific pathway, a powerful channel for *intuitive* anti-Blackness to be performed. The imagined naturalness of intuition, the cherished egalitarianism of play, and the libidinal relief assumed to come with humor make improv particularly anti-Black through its aesthetic specificities.

This working hypothesis has crystallized for my project: improv is a ground, a *safe space*, for anti-Blackness to be realized under the rhetorical veil of progress, freedom, democracy – all of which have themselves been enabled and catalyzed qua anti-Blackness. This differentiates improv from other forms of expression in its parasitic procedure of spontaneous reenactment. Improv is therefore a case worthy of particular consideration. While it is feasible to argue that the world is flawed and that improv's flaws are a direct consequence of that (if understood in its specific configuration of intuition, humor, and play), improv's discursive design creates a terrain for the individual and collective libidinal economy that structures white modernity. Intuition, play, and humor can (and should) all be theorized in their libidinal motivation in the specific ways that empower and propel them, that *drive* them. Intuition, play, and humor can be linked conceptually to the anti-Black affect of modernity in the language of psychoanalysis. Improv, relying as heavily as it does on *the Human* who plays and laughs, must be affected by a critical examination of the concept.

3 Truths for Whiteness

3.1 Sylvia Wynter's conception of being hu/man

Adapting truths: historicizing humanism

In the introduction, I suggest that the three investigative frames of this project – intuition, play, and humor – all mobilize some notion of "the human" in theory, practice, and analysis. Paul Sills, for example, states: "I think improvisational theater and my mother's work are attempts […] to go into the possibilities of human development" (qtd. in Sweet 17), and that "theater is responsible for the image of the human [and] the concerns of the artists are the concerns of the people" (qtd. in Sweet 18). Who or what constitutes "the human" or "the people" in question here? The project of human self-definition gains traction through intuition. Intuition plays a central role in any attempt to overcome or transcend consciousness, whether in religious, artistic, mysticist, or esoteric practices, or in contemporary scholarship on cognition and decision-making. Somehow, through intuition, we can allegedly heal ourselves or develop a more perfect, spiritual existence by way of integrating *something else* into our conscious being in the world. So too with humor: animals may cackle or laugh, but only Humans have humor. Aristotle's *homo ridens* still features heavily in introductions to humor theory. And lastly, of course, hardly any academic treatment of "play" can do without Turner's *homo ludens*, or Schiller's dictum that "man only plays when in the full meaning of the word he is a man, and he is only completely a man when he plays" (80). In this chapter, I look into the historical and discursive development of this universalized and universalizing Humanism. I lay out where our contemporary idea of modern Humanity comes from and how it hides its contingency and cultural specificity from white subjects by presenting itself as spiritually universal and objectively absolute.[1]

In "Unsettling the Coloniality of Being/Power/Truth/Freedom," Sylvia Wynter writes that our "varying ontogeny/sociogeny modes of being human, as inscri-

---

[1] Note that, in this sentence, the referent-we functions generically, partaking in a discourse that creates the illusion of such genericity in the first instance. I maintain the use of we, us, or ourselves not to assume a shared positioning of myself and the reader, but to mark myself as within an episteme that cannot be transcended by the stroke of a hand. By doing so, I hope to mark my own position as an inhabitant of the world, in which "we" refers to the culturally specific realm of subjectivity. The "us" in this project is always culturally specific.

bed in the terms of each culture's descriptive statement, will necessarily give rise to their respective modalities of adaptive truths-for, or epistemes, up to and including our contemporary own" (269). Any *truth* about humanity is never absolute but always a "truth-for," a cultural episteme emerging from a descriptive statement provided by a certain discourse around what it means to be human at a given time in a given place. Modern European imperial enterprises required a conceptual shift in what it meant to be human because the theocentric descriptive statements from the Middle Ages were challenged by colonial conquest. The medieval truths-for that defined human subjectivity as being a good Christian had to be modified to lend moral justification and political legitimacy to the expropriation of what was called the "New World" and the centuries of mass enslavement that have since defined modernity. Wynter argues with Aníbal Quijano that within the modern European political and military endeavors, "'the idea of race' would come to be the most efficient instrument of social domination invented in the last 500 years" (Wynter 263). However, the concepts of race and Blackness have not always been the signifiers for the social, epistemological, and political construct that I refer to as a "matrix slot of Otherness," in Wynter's phrase (266). In medieval times, the matrix slot of Otherness was applied to the enemies of Christ. The Christian subject, the Human before Humanism, was created in the image of God, and the infidel was its Other, "with Jews serving as the boundary-transgressive 'name of what is evil' figures, stigmatized as Christ-killing deicides" (Wynter 265–66). Whereas Christianity defined what it meant to be human in religious and theocentric terms, modernity introduced a descriptive statement on the grounds and in terms of rationality and politics. This is the emergence of Humanist Man1, as Wynter terms the political subject as the first variant of the modern Human.[2]

Wynter claims that this shift from the medieval Christian subject to the ratio-political Man of modernity did not involve a complete epistemological overthrow of the Christian dichotomy of Spirit/Flesh (in which the clergy stands at the top and the laity at the bottom). Rather it *transumed*[3] this binary – a reformulation,

---

[2] The numerical qualifier of Man1 serves later to differentiate this homo politicus from the later – biocentric – variant of the modern Human, Man2.

[3] Wynter adopts this term from Harold Bloom, who notes that "cultural fields are kept in being by transumptive chains" (qtd. in Wynter 308). She mobilizes it to refer to a "carrying over" of tropes through different historical contexts in order to "keep the community going by means of its retroping of earlier tropes," quoting Bloom in "How We Mistook the Map for the Territory" (2016). In this project, I rely on it heavily as an argumentational tool, which I need not only to argue in historical diachronicity but transversally through different disciplines as well as linguis-

both historically adaptive and culturally specific, of what the Middle Ages had discursively inherited from the descriptive statement of ancient Greek society. For the Greeks, the descriptive statement of what it meant to be human was mapped onto the physical cosmos, which served to create the illusion of "supernaturally (and, as such, extrahumanly) determined criteria" for human subjectivity: "this value distinction (sociogenic principle or master code of symbolic life and death) then being replicated at the level of intra-Greek society, in gendered terms, as between males, who were citizens, and women, who were their dependents" (Wynter 271–72). An assumed extra-human agency was understood to have organized the world in a specific way beyond human control. Even though this agency was constructed by the subjects of Greek society and only subsequently imposed upon the stars, for the subjects who lived within that order, this dichotomy was

> the indispensable condition of their existence as such a society, as such people, as such a mode of being human [that] commanded obedience and necessitated the individual and collective behaviors by means of which each such order and its mode of being human were brought into existence, produced and stably reproduced. (271)

In understanding the arbitrariness of Greek self-description, we can better understand the contingency as opposed to the apparently "natural" or "God-given" status of our own. Such truths-for were never as universally absolute as they presented themselves to the subjects and the societies in question but always "remained adaptive truths-for" (271). Judeo-Christian Europe inherited the axiomatic Platonic postulate "of an eternal, 'divinized' cosmos as contrasted with the Earth, which was not only subject to change and corruption but was fixed and unmoving at the center" and reframed it in its specifically Christian terms. The theocentric statement of the Middle Ages was thus a transumption of the Hellenistic "master code" and similarly mapped the perfection/imperfection concept onto the heavens and the earth, only this time in its conception of the Adamic Fall as significant sin (271–72). Taking her cue from David Bohm – who in a 1987 interview points out that "each general notion of the world contains within it a specific idea of order," and that the specific order of the ancient Greeks was based on "the idea of an increasing perfection from the earth to the heavens" – Wynter argues this idea of order was, by way of transumption or mapping, "Christianized within the terms of Judeo-Christianity's new 'descriptive statement'" of human being, based on its master code of the 'Redeemed Spirit' (as

---

tic and affective registers. Transumption, transuming, and transumptive are helpful terms for theorizing in a sphere where superficial decoding is often difficult to perform.

actualized in the celibate clergy) and the 'Fallen Flesh' enslaved to the negative legacy of Adamic Original Sin, as actualized by laymen and women" ("Unsettling" 274).

However, the making of modernity following the Middle Ages was not as radical a revolution as is often suggested, but a smooth transition that was fully operational within the transumptive chain. While the new *bourgeois king* was no longer officially appointed by God but by the allegedly objective concept of rationality – and the political rationale of the modern West of course – modern Man as the subject of Humanism had to argue from *within* theocentric logic. The lay intelligentsia of modernity, finding "themselves in a situation in whose context, in order to be learned and accomplished scholars, they had had to be accomplices in the production of a 'politics of truth' that subordinated their own lay world and its perspective to that of the Church and of the clergy" (276), developed a Humanist concept within theological terms, creating a "hybridly religio-secular" version of Man that drew on the residue of the theocentric descriptive statement as well as the budding autonomy of the modern subject. Humanists successfully shifted away from the absolutism of the Church and its (previously condemned) earthly world by arguing that God had made Man in his model to admire him, and then left him "to decide for himself whether to fall to the level of beasts by giving in to his passions, or, through the use of his reason, to rise to the level of angels" (276–77). It was believed that, though responsible for its intelligent design, "He would have had to make it according to rational, nonarbitrary rules that could be knowable by the being that He had made it for" (278). This reformulation made it possible for a Human subject to emerge, Man1, which could be circumscribed without reference to Christian terminology as "the rational political subject of the state, as one who displayed his reason by primarily adhering to the laws of the state" (277). Reason and political law re-enforcing one another, the power no longer lay with the priest, but with the constitution; lawful behavior was substituted for piety. Existence was now conceptualized through a Humanist lens. A ratio-political descriptive statement superseded the theocentric one and, drawing on the transumptive chain, provided a new map for a preexistent discursive territory:

> Spiritual perfection/imperfection, an idea of order centered on the Church, was now to be replaced by a new one based upon degrees of rational perfection/imperfection. And this was to be the new "idea of order" on whose basis the coloniality of being [...] was to be brought into existence as *the foundational basis of modernity*. (287–88)

Given the emergence of the modern matrix of discursive Human existence, what it meant to be human had to be re-negotiated terms different from theological compliance or conversion.[4] No longer did "heretics [...] Enemies-of-Christ infidels and pagan-idolators" serve as "the physical referents of the conception of the Untrue Other to the True Christian Self." The "matrix slot of Otherness" was vacant. There was yet no reassuring (b)order that defined what the Western European subject of modernity was not ("Unsettling" 265–67).

With the wisdom of hindsight, we must observe that modern Man did not realize the full potential of disposing with the theocentric descriptive statement. Hypothetically, this was a historical moment when human existence could have been rethought entirely without reliance on or recurrence to the dichotomies of Heaven and Earth, Christian and non-Christian, instead developing an idea of humanity that involved no fundamental *ex negativo*. However, because of the discursive logic grounded in the configuration of political power still held by Christianity, dehumanization and violence marked the beginnings of modernity: Man invented "race" as a solution to this crisis in self-definition to fill the matrix slot of Otherness. The construct of race "was therefore to be, in effect, the non-supernatural but no less extrahuman ground [...] of the answer that the secularizing West would now give to the Heideggerian question as to the who, and the what we are" (264). Accordingly, race also answered the question as to who and what "we" are not: "the peoples of the militarily expropriated New World territories (i.e., Indians), as well as the enslaved peoples of Black Africa (i.e., Negroes)" (266). Humanism as a European or Western self-construction cannot be separated from the dehumanization of its colonized and enslaved Others,[5] and Western modernity realized and actualized its new descriptive statement through colonialist and enslavist enterprises. In a "second wave of imperial ex-

---

4 Wynter elaborates this negotiation in detail in discussing the argument between Las Casas and Sepulveda:
> a dispute that I will define as one between two descriptive statements of the human: one for which the expansion of the Spanish state was envisaged as a function of the Christian evangelizing mission, the Other [sic] for which the latter mission was seen as a function of the imperial expansion of the state; a dispute, then between the theocentric conception of the human, Christian, and the new humanist and ratiocentric conception of the human, Man2 [sic, probably Man1] (i.e., as homo politicus, or the political subject of the state). (269)

5 The concepts of Others or Othering, as I understand them, are part of a register of relationality that does not hold up in contemporary usage when we move towards anti-Black, abjective modernity, as I will show. I retain the term here to stay with the language of Wynter and work with her conception of the matrix slot of Otherness. She mobilizes this phrase to conceive of a specific Black Other that is so absolute and ultimate in its dehumanized (and abject) existence that relationality ultimately no longer applies. I also find the term useful to mark that modernity's creation of and reliance on anti-Blackness is culturally specific, even in the radicalness of this specific Other's manifestation.

pansion," the West further transumed Man1 in biological, Darwinian terms as Man2 (266). This time "it was to be the peoples of Black African descent who would be constructed as the ultimate referent of the 'racially inferior' Human Other" as the "negation of the generic 'normal humanness,' ostensibly expressed by and embodied in the peoples of the West" (266). Among various Others such as the Savage, the most salient "was to be that of the mythology of the Black Other of sub-Saharan Africans (and their Diaspora descendants)," whose systemic stigmatization, social inferiorization, and dynamically produced material deprivation served to "'verify' the overrepresentation of Man as if it were the human" (267). Drawing on the Ancient dualism of heaven and earth, this was "done in a lawlike manner through the systemic stigmatization of the Earth in terms of its being made of a 'vile and base matter,' a matter ontologically different from that which attested to the perfection of the heaven" ("Unsettling" 267). Blackness was constructed as the ultimate and absolute Otherness of Humanity.

What differentiates the modern descriptive statement from the previous one was the radical way in which Man construed himself as universally true and objectively absolute:

> While, as Christians, Westerners could see other peoples as also having gods (even if, for them, necessarily "false" ones as contrasted with their "true" and single One), as subjects defined by the identity Man, this could no longer be the case. Seeing that once its "descriptive statement" had been instituted as the only, universally applicable mode of being human, they would remain unable from then on until today, of [...] conceiving an Other to what they call human. (299)

The result, as Wynter proposes, was "that the new master code of the bourgeoisie and of its ethnoclass [biocentric] conception of the human [as Man2] was now to be mapped and anchored on the only available 'objective set of facts' that remained," namely "the set of environmentally, climatically determined phenotypical differences between human hereditary variations" (315). Because God was no longer understood to be in charge of societal order, that order had to be *scientifically observed and then performatively actualized*. Supernatural causation was substituted for natural causation, which was used to legitimize and justify mass enslavements and genocides (304). The belief in objectivity and the development of the physical sciences served perfectly to "make opaque to themselves/ourselves [...] the empirical fact of our ongoing production and reproduction of our order, of its genre of being human, its mode of consciousness or mind, and therefore of the latter's adaptive truth-for" (307). Wynter here builds on Godelier's analysis of the "mechanisms that function to project their/our authorship [of the order we live in] onto Imaginary supernatural Beings" or their transumption. As in Ancient Greece, an observed reality is rendered abstract as the ostensible realization of a higher ontological order. The active maintenance of the matrix slot of Otherness still serves to veil "our own authorship and agency" of and in today's white, modern order ("Unsettling" 315). It allows us, as white subjects of that modern order, to construct ourselves as *normal* Humans with a perceived phenotypical difference from those who do not have the capacity for this same subjectivity. With Darwin, this discursive matrix gave rise to another manifestation of the subject-making dichotomy: those selected and those dysselected. The Darwinian, naturally selected Man2 is the manifestation of a culturally specific discursive design for the way in which white subjects articulate "biological Blackness" through the registers of cultural ascription or in economic terms. We are now in a space where our *belief* in naturally given criteria and their objective categorization makes it almost impossible to conceive of the fact that this is merely a culturally specific descriptive statement. Who wants to argue with such facts?

The killing of God's creation – a paradoxical act in Christian terms – became possible through the shift towards the ratiocentric Human. The matrix slot of Otherness was still the same slot for the same people because the referent of normalcy was its European same: "the Elect category of those redeemed from [...] sin has now been recast in terms of the 'by-nature difference' of rationality" (304). Within European imperialism, Black enslavement and genocides kept morality intact while simultaneously and reciprocally demanding from the white, modern subject a psycho-epistemological strategy that allowed its actions to align with Humanism. Our contemporary descriptive framework of biocentric Man2 (as transumed from the political subject Man1 as transumed from the Christian subject) is necessarily still "inscribed within the framework of a specific secularizing reformulation of that matrix Judeo-Christian Grand Narrative" (318) – and anti-Blackness is the governing structural ground within that preexistent matrix. Race, then, must be understood as *a structuring principle and not a feature* or avenue for discrimination. Rather than a ground for negative stereotypes, race "was and is fundamentally the issue of the genre of the [modern] human" ("Unsettling" 288). It is impossible to disentangle whatever is formulated within the Humanist matrix from its anti-Blackness, because anti-Blackness provides the negation of the subject within the logic of that originary dichotomy. Without recognizing the function of anti-Blackness, we can make no sense of the Human as we know it today. While still largely ignored in mainstream cultural (or other) studies, this is consequential for the Humanities at large. Anything done in the name of Humanism, any discipline that focuses on the Human, any object (or its study) that mobilizes Humanist ideas is always already inseparably bound up with the racialization and dehumanization of Blackness and those whom the discourse marks as such. Wherever "the Human" is mobilized and built upon as a given, sensitive scrutiny is required.

Being human in bios-mythoi hybridity

I hinted above that when modernity transumed the theocentric descriptive statement of the human from the medieval Spirit/Flesh dichotomy, the Christian subject could have been overthrown completely if being-human-in-dichotomy had been called into question. If the subject could be thought outside of such a dualist (or binary) matrix, being human could create subjects that need no Other to make themselves, as Wynter suggests in her early essay "Ethno or Socio Poetics" (85). This would demarcate an actual rupture of the transumptive chain, and demand that we think about human being in an entirely different framework. It would need an axiomatic ground in which human being is neither static nor separated into categories like Spirit/Flesh (Christian subject), rational/irrational (political Man1) or selected/dysselected (biological Man2). Wynter has set out to develop such a conception. In "Toward the Sociogenic Principle," she takes a cue from Fanon's postulate that "[a]llongside phylogeny and ontogeny, there is also sociogeny" (Fanon 4). The sociogenic principle is central in Wynter's critique of "our present culture's purely biological definition of what it is to be, and therefore of what it is like to be, human" ("Principle" 31). Fanon's sociogeny builds on Freud, who took the terms ontogeny (the study of an organism) and phylogeny (the study of a species) from Ernst Häckel's recapitulation theory ("ontogeny recapitulates phylogeny"), which assumed that an individual organism undergoes the same evolutionary phases as its species. Häckel was one of the first to conceive of *politics as applied biology*. He was a forerunner of eugenics and was enthusiastically received in Nazi Germany; Freud draws on this background when he conceptualizes the terms ontogeny and phylogeny for psychoanalysis. Introducing sociogeny, Fanon disposes of the naturalized, racial essentialism previously inscribed in the ontogeny-phylogeny relationship as developed by Häckel and Freud, while maintaining the axiom that the individual lives and exists *by* and *through* the species. Fanon rids the concepts of their "natural" and materialist determination while acknowledging the role of physiology of the human organism. With Fanon, we can think of the individual human body as a physiological entity that is nonetheless actually and effectively (in)formed by "a symbolic register, consisting of discourse, language, culture, and so on," which "accompanies the genetic dimension of human action" (Wehelye 25). For Wynter, as Alexander Wehelye notes, this provides a way to think being human in a space "where culture and biology are not only not opposed to each other but in which their chemistry discharges mutually beneficial insights" (25). Sociogeny provides a sphere to theorize being human as a praxis departing from the level of the socially performative. It offers an entirely different conception of what it means to be human, one that directly addresses the human body in its biochemical existence without resorting to determinist materialism or giving in to semiotic or performative musing. Sociogeny involves an entirely novel conception of being human.

Wynter elaborates in various essays and interviews on her idea that "the human is, meta-Darwinianly, a hybrid being, both *bios* and *logos* (or, as I have recently come to redefine it, *bios* and *mythoi*" ("Catastrophe" 16). She always makes clear that "should this hypothesis prove to be true, our system of knowledge as we have it *now*, goes." The transumptive chain discussed above, which has always forced theorizers to ground human subjectivity discursively on the master codes of Spirit/Fallen Flesh, would be massively disrupted. The entire system of our understanding of the world, and of ourselves, would break apart: "If human beings are conceptualized as hybrid beings, you can no longer classify human individuals as well as human groups, as *naturally selected* (i.e., eugenic) and *naturally dysselected* (i.e., dysgenic) beings. This goes away. It is no longer meaningful" (17). Wynter suggests that we, as *homo narrans, as* being human in *bios/mythoi* hybridity, came into existence by way of a Third Event:

> The First and Second Events are the origin of the universe and the explosion of all forms of biological life, respectively. I identify the Third Event in Fanonian-adapted terms as the origin of the human as a hybrid-auto-instituting-language-storytelling species: bios/mythoi. The Third Event is defined by the singularity of the co-evolution of the human brain with – and, unlike those of all the other primates, with it alone – the emergent faculties of language, storytelling. This co-evolution must be understood concomitantly with the uniquely mythmaking region of the human brain, as the brain scientists Andrew Newberg, Eugene D'Aquili and Vince Rause document. ("Catastrophe" 25)

Locating her conceptualization of being human in the discipline and language of neuroscience, she defines the sociogenic principle as

> the information-encoding organizational principle of each culture's criterion of being/nonbeing, that functions to *artificially* activate the neurochemistry of the reward and punishment pathway, doing so in the terms needed to institute human subjects as culture-specific and thereby verbally defined, if physiologically implemented mode of being and *sense of self*. ("Principle" 54)

Because the symbolic triggers for affective procedures that "activate the neurochemistry" are artificial, there is one central consequence: "*humanness* is no longer a *noun. Being human is a praxis*" ("Catastrophe" 23). If we had other criteria instead of the dichotomies of the transumptive chain that result in the matrix slot of absolute Black Otherness, our brains would look different, would be activated by different triggers. We would feel different, experience the world in entirely different and yet unimaginable modalities. Being human, then, is a continual act of reasserting the bodily integrity of our sense of self, which is symbolic nonetheless. The simultaneity of *bios* and *mythoi* (not opposed to but along with each other) in being human-as-praxis is the theoretical sphere in which this project is located.

Psychoanalysis as embodied cognition

Though starting from Fanon, Wynter's conception departs from his in that she is not interested in psychoanalysis or any of its tools. Instead, she frames her bios/mythoi concept not only as decidedly meta-Darwinian but also meta-Freudian ("Catastrophe" 54). As I build largely on Wynter, why would I mobilize the psychoanalytic toolkit? I find it methodologically helpful because, ultimately, this project needs a language that communicates with the discourse. This is not a self-aggrandizing attempt to *further advance* or *develop* Wynter's theorization of being human, and so I seek to work from her axioms while drawing on a conceptual framework with which I can immediately address the fields in question: intuition, play, and humor as realized in improv. This also helps to frame Broeck's notion of *anti-Black abjection* in the *culturally specific* libidinal economy of white modernity as the praxis of maintaining the white modern subject qua the continual abjection of Blackness – that is, the dehumanization of any body, object, or phenomenon racialized as Black. This in turn resonates strongly with Wynter's definition of the sociogenic principle as presented above, even though articulated in the language of psychoanalysis. The psychoanalytic concepts I apply always remain tethered to original and contemporary findings and positions developed in embodied cognition theory, which Wynter also uses to ground her theory. Currently, and despite the celebration of affect theory, the disciplines and languages of psychoanalysis and embodied cognition do not regularly communicate as productively as they could. Indeed, thinking psychoanalysis alongside embodied cognition runs the risk of getting ground up between the ideologies of the *black box*[6] of the unconscious, and that of an elim-

---

6 Among the many phrases in which common usage of the adjective *black* manifests white modern racialized knowledge – for example, "black humor" being concerned with sexuality or death, and "*blacking* out" denoting a loss of consciousness and the reduction to automatic bodily functions – black box is particularly telling, because we can only see the outside or the sur-

inative materialism. While cultural studies scholars are starting to speculate on the potential of their integration, representatives of the respective disciplines do not have much to say to each other. Consider cognitive scientist Guy Claxton, who addresses the interdisciplinary relationship directly in *Intelligence in the Flesh:*

> At my most radical, I would now claim that, not only are "the gods and spirits" non-existent (even though they may still have their uses), but the unconscious is dead too. We may choose to continue using it as a metaphorical or poetic way of talking but there ain't no such animal. There are myriad processes in the body that never lead to conscious experience, but there is no real, identifiable place or agent inside us that is a separate source of impetus from consciousness and reason. Like "the mind," "the unconscious" is a place-saver, a dummy explanation. It is like a temporary filling in a tooth, put there till something better comes along. And now it has. (12)

But has it? Freud may well have subscribed to this. The adoption of terms like ontogeny or phylogeny from biology speaks to the way in which he framed his theorizations as natural science. In "The claims of psycho-analysis to scientific interest," he articulates the relevance of psychoanalysis for various disciplines, including biology and the language sciences, which, if properly framed, would make psychoanalysis an apt instrument for the study of being human as bios/mythoi. In "Beyond the Pleasure Principle," Freud speculates about a "specific locus" for consciousness by "aligning ourselves with the locational hypotheses of cerebral anatomy" (64). Kathryn Armistead speaks for many scholars critical of this version of Freud, who "[w]herever possible [...] substantiated his conclusions using arguments based on physical, biological reality." She understands these as strategic gestures to ensure "his own place in the history" at a time when the natural sciences were in the march (2). Be that as it may, there is something to be recognized in the relationship of early psychoanalysis with the natural science of what would later be called cognitive science. For Freud, "the mind, whether conceived of as conscious/unconscious or ego/superego/id was always a product of the brain," which relates to contemporary theories of embodied cognition and their questions about consciousness. If Freud was convinced that when therapists and patients unearthed "mental artifacts" in the analytic process, this would cause a "material difference in the patient" that also relates to cognitive science (Armistead 2). If nothing else, this does give us a cue to go onward.

---

face of that box. We can control what goes in, and we can see what comes out. But what goes on in there in empirical actuality, white modern science has given up on knowing.

In this project, I attempt to think embodied cognition and psychoanalysis together, each informing the other, seeking to develop an understanding of some select concepts that ultimately feed back into one another. As suggested by Claxton, I will use the language of psychoanalysis as a "place-saver" or "dummy explanation" for what goes on in the body. However, I do so not in ignorance of the idea that "something better has come along," but because cognitive science only does half the trick when it comes to the conception of being human per Wynterian bios/mythoi hybridity. Working in a psychoanalytic framework and applying the language of an unconscious *structured like a language*, as suggested by Lacan, does not at all necessitate the disposal of an individual organism's biochemistry. The latter feeds into and substantializes the former, and the former can help us to talk about and interpret the latter.[7] And while neuroscience helps us to realize and observe the mechanics, the psychoanalytic instrumentarium can do what it does best: labor at developing a language for an object that is absent. A language for something that only shows itself in its effects but not in its ontology – because it has none. The psychoanalytic object of study is something that by definition we cannot know, but which provides meaning to what we are and do. Neuroscience cannot do that. Not only do I think it is valuable to keep working with psychoanalytic concepts "as a convenient shorthand for talking about complex human processes and experiences" alongside the evidence put forward by experiments of and for embodied cognition to rethink being human in Wynter's vein (Claxton 139). I also build from Claxton's admission that "[s]ome of what is interesting about us humans can be well talked-

---

[7] Lyotard's *Libidinal Economy* suggests this perspective as well by mobilizing the body as both metaphor and actual space for the goings-on of what Freud conceived of as libidinal economy. The book opens with a vivid and literal "opening" of a body, involving all biological matter imaginable, then develops the metaphor into the Moebius strip on which linguistic representation takes place, in the "theatrical volume," which then provides the language and framework for the libidinal economy. In this way, Lyotard can conceive of libidinal economy as being one with the actual body:
> Don't forget to add to the tongue and all the pieces of the vocal apparatus, all the sounds of which they are capable, and moreover, the whole selective network of sounds, that is, the phonological system, for this too belongs to the libidinal "body," like colours that must be added to retinas, like certain particles to the epidermis, like some particularly favoured smells to the nasal cavities, like preferred words and syntaxes to the mouths which utter them and to the hands which write them. (2)

With Lyotard, we can say that "libidinal economy" is a way to conceptualize the body as the primary locus not only of hunger, thirst, and digestion, but also of feeling, thinking, imagining, and as matter that itself is capable of learning. This corresponds well with Wynter's demand for the recognition of being human as a bios/mythoi hybrid praxis, thinking the symbolic and the biological alongside each other.

about in body terms, while other aspects are handled more elegantly by mind language" (140). Some aspects of being human will come to the fore in one discipline, others in the other, and some precisely in the space between, opened up by psychoanalyst Fanon's conception of sociogeny.

3.2 The modern libidinal praxis of anti-Black abjection

Libidinal economy of Black enslavement

In *Red, White & Black*, Frank Wilderson propagates a "radical return to Fanon." Unlike Wynter, however, he elaborates on Fanon's psychoanalytical dimension, taking into view the "libidinal economy of civil society" that structures and drives the psychosociality of white modernity (15). He defines it with Jared Sexton as "the economy, or distribution and arrangement, of desire and identification (their condensation and displacement), and the complex relationship between sexuality and the unconscious" (qtd. in Wilderson, *Red* 49). Building on Orlando Patterson, who defines the social ontology of enslaved people as one of social death, Wilderson argues that slavery in modernity has become "the singular purview of the Black," who has been "socially dead in relation to the rest of the world" ever since. Where in Patterson's work enslavement is understood as "a condition that anyone can be subjected to" (18), for Wilderson "slavery is and connotes an ontological status for Blackness" (14), providing the only "access to (or, more correctly, banishment from) ontology" for those racialized as Black. They are thus "by definition always already void of relationality" (18). The primary function of the construction of Blackness is, as Wilderson argues, symbolic rather than economic. He follows Eltis in assuming that it would have been much more profitable to seize "convicts, prisoners and vagrants" from Europe, rather than taking up the financial effort of sailing to Africa (15). Granting that "the constituent elements of slavery are not exploitation and alienation [as Patterson suggests], but accumulation and fungibility (as Hartman puts it): the condition of being owned and traded" (Wilderson 14), the enslavement of white people was unthinkable for white people. Enslaving a fellow white subject, even if they might have had a life sentence in prison, would have "stripped the convict of the aura of the social contract" (15). Such social autoaggression or autodestruction was unimaginable because it would have challenged the anti-Black system at large. Wilderson writes that in an imagined process of white-on-white enslavement, "[w]hat Whites would have gained in economic value, they would have lost in symbolic value; and it is the latter which structures the libidinal economy of civil society" (15). Further, if people racialized as

3.2 The modern libidinal praxis of anti-Black abjection — 77

Black were understood as agents with cognitive, rational powers, like their white counterparts, the absence of legitimation of their subjection would be overt:

> The race of Humanism (White, Asian, South Asian, and Arab) could not have produced itself without the simultaneous production of that walking destruction which became known as the Black. Put another way, through chattel slavery the world gave birth and coherence to both its joys of domesticity and to its struggles of political discontent; and with these joys and struggles the Human was born, but not before it murdered the Black, forging a symbiosis between the political ontology of Humanity and the social death of Blacks. (Wilderson 20–21)

Blackness served as the referential background of the semiotic chaos, against which free Europeans could cut out their own image of civilized subjectivity – both in discursive and affective terms.

I state this explicitly because I suggest that Wilderson's theorization of the anti-Black libidinal economy can and should be related to Wynter's concept of the matrix slot of Otherness. I am aware that the term "Other" can be understood as implying relationality, which – speaking with Wilderson – never encompasses Blackness. Yet I read Wynter's understanding of Man in its absolute universalization of the white self in that very non-relation to Blackness as developed by Wilderson. As mentioned above, modernity brought about the culturally specific transumption that there were, in fact, no Others to the white subject of Man, "overrepresented as the generic, ostensibly supracultural human" (288); for the biocentrically described Man2, Blackness served as the "missing link" connecting "true (because rational) humans and the irrational figure of the ape" (Wynter, "Unsettling" 304). This allegedly naturally determined existence outside Humanity has no relationality, no discursive potential for political subjectivity or political ontology, to use Wilderson's term. Similarly, Wilderson emphasizes the libidinal dependence of whiteness on the existence and active maintenance of Blackness. Just as Hobbes could not articulate his concept of individual subjectivity in terms of self-possession without imagining bodies lacking such self-possession, there can be no freedom without relation to the most *naturally unfree*, which is the figure of the Black: "Without the Negro, capacity itself is incoherent, uncertain at best even" (Wilderson, *Red* 15). Anti-Blackness is the ground on which possibility itself can be played out on any kind of sociopolitical grid – which does not imply that every position is of power, but that every position is of potential power. Resonating with Wynter, Wilderson applies the term "matrix" to address the existential dependence of white subjective existence on the social death of Black-racialized people:

> Whiteness is parasitic because it monumentalizes its subjective capacity, its lush cartography, in direct proportion to the wasteland of Black incapacity. We should think of it as a kind of facility or matrix, through which possibility itself can be elaborated. (*Red* 45)

I suggest, then, that with Wilderson we can *libidinalize* Wynter's notion of the matrix slot of non-Human Otherness as a sphere of Blackness that denotes non-capacity, non-being, and non-Humanity. It is a space flexible enough to incorporate every *non-* required by any given time or place. The (hypothetical) recognition of Black capacity or agency would require the complete disintegration of current formations of sociality because the entire affective grammar through which the white modern subject knows itself would lose its grip. There is no alternative to the stably reproducible "physiologically implemented mode of being and *sense of self*," as Wynter writes in her definition of the sociogenic principle ("Principle" 54).

It must be understood that Blackness is not a by-product of white subject-making but the foundational structure of the universalized white self of modernity for whom it is otherwise impossible to self-conceptualize "ex positivo." Given the modern descriptive statement as its discursive prefiguration, the white subject relies on Blackness *ex negativo* and on the continual performance of anti-Blackness, because "with being human everything is praxis" (Wynter, "Catastrophe" 34). In what follows, I elaborate on the notion of anti-Black abjection developed by Broeck as the praxis central to the making of the modern white subject. I read anti-Black abjection as a culturally specific practice performed and observed on many levels, including its visceral and neurochemical hardwiring. I hope to formulate some of the dimensions of this praxis across all aspects of white existence, from actions to perceptions, from political actions to musical tastes. I will focus on the libidinal ground, which I posit as the primary motivational force for all things white as expressed in myriad cultural and artistic practices.

Abjection with Broeck and Kristeva

Broeck coins the phrase "white Euro-American abjectorship" ("Legacies" 109), reading abjection "in the post-Fanonian vein in which it appears in Saidiya Hartman's and Hortense Spiller[s]'s work [as] a theoretical concept to discuss the underside of white Western modernity's terms of human sociability and subjectivity" (*Gender* 13). Departing from Kristeva's treatment of the psychoanalytic concept, Broeck does not concern herself with a phenomenology of the abject in its alleged irreducibility or its potential to disturb the symbolic order but mobilizes its predicative performance of white subject-making grounded in the dehumanizing practice of enslavement. She writes:

> It is important to stress, though, that, while I am borrowing from Kristeva's notion of the abject as that which threatens the subject's secure anchoring in the symbolic from an elsewhere, I am in contradiction to Kristeva. I am not interested in her question of what the abject does to an individual or collective subject – plunge it into states of disorientation – but in the way in which the white modern subject (male and female) might be considered an abjector, that is, a motorizing force which needed Black thingification to "know" socially, culturally, politically and epistemically, its subjectivity and social being. (*Gender* 17)

Broeck thus mobilizes the term with Hartman "to be able to talk about the positioning of human beings as female flesh, as that abject which has been most radically beyond the pale of the subject in an Enlightenment vision" (*Gender* 13), as a performative action that ensures the "bounded bodily integrity of whiteness secured by the abjection of others" (Hartman, *Scenes* 123). This historicized usage of the term is helpful to discuss what up until today "has been structurally, not contingently, cut off from the human, from the self-possessed possessor of the world and its things" (*Gender* 13). White abjectorship for Broeck describes a system and the predicative transitivity of white subject-making that encompasses both internal (as a psychic, self-making principle) and external (directed against others) activity. In a systemically anti-Black abjective structure, a white body attains subject status qua *acts* of outward anti-Black abjection and can also experience psychophysiological sensations of *inward* acts, as I suggest below.[8] In the modern system of white abjectorship, subjectivity without anti-Black abjection is inconceivable. There is no white subject who is not an active anti-Black abjector, because we can neither escape a) the transumptive chain that prefigures us to conceive of ourselves in descriptive statements based on dichotomies, nor b) the cultural specificity of our present descriptive statement, which is semantically realized through anti-Blackness. In this inward libidinal process, the psychic space of "Blackness" is created, a space without limits that cannot be determined positively. White abjectorship thus denotes the (mode of) *doing whiteness* qua culturally specific anti-Black abjection, creating a boundless sphere of the white non-: non-subjectivity, non-sociability, non-being. This discursive non-space is central to the sociogeny of the modern West, informing its phylogeny and ontogenies. Individual performances of anti-Black abjection correspond to the continual performative making of white sociability as such, which according-

---

[8] The notion of anti-Black abjection through "inward acts" likely stretches Broeck's use of the term.

ly holds subjective capacity for whites only. In order for the system of white abjectorship to work, any body, object, or phenomenon cast (off) as "Black" is treated in the same predicative manner as non-relation to the reigning subject as a visible performance of white abjectorship.

Where for Kristeva the subject is "*beset* by abjection" (2, emphasis mine), Broeck deploys the concept as something *done* rather than *experienced*. She conceives of the abjective process as an action grounded in the psychic social and individual anti-Black structure described above. Anti-Black abjection thus describes not only a systemic order of the modern West but also its repeated reinstatement, its continually performed actualization in the real world, the singular acts and performances enabled by and performed for the anti-Black systemic order. Through anti-Black abjection-as-action, the white subject reinstalls itself when- and wherever necessary – or simply because it has the capacity to do so in a leisurely pursuit of jouissance. What can be gained from this reconsideration of Kristeva? First, given the libidinal dimension of anti-Black abjection in Broeck's and Hartman's use of the term, a return to Kristeva can indeed offer insight because culturally specific semantics are always read along and mapped onto her argument. When read through the culturally specific and systemic anti-Black actualizations of abjection, many of the terms and phrases in her detailed description of abjection reveal how naturalized concepts like disgust must be understood within the anti-Black discursive matrix. Reading Broeck alongside Kristeva is thus metatheoretically relevant. Second, I find a small conceptual treasure in Kristeva's essay, which – at least by way of analogy – resonates with other concepts in the psychoanalytic toolkit I draw on in discussing intuition, play, and humor. This immediate connectivity is worth maintaining, as long as we keep cultural specificity in mind and are never seduced by a universalized assumption of generic humanity. This gesture is inspired by Mills's argument for a mobilization of "contract talk" to elaborate on white supremacy, suggesting that the idea of a Racial Contract is "one possible way of making this connection with mainstream theory, since it uses the vocabulary and apparatus already developed for contractarianism to map this unacknowledged system" (*Contract* 3). To draw on Kristeva's psychoanalytic classic has its advantages, even though the self-propelling logic of "unconscious talk" must at times be tethered to the cultural specificity and actual cultural phenomenon in question. Subsequently, I will mobilize concepts of embodied cognition to suggest we can understand anti-Black abjection as a definable somatic state that provides us white subjects with a very special jouissance of revamped symbolic and bodily integrity. In doing so, I hope to contribute to a "thick description" that "understand[s] the intricate psychic, social, and intellectual mechanics of European modernity's culture of self as ownership, the mechanics of abjectorship" (Broeck, "Legacies" 121).

Jouissance of "I"

Kristeva defines abjection as an instinctive and pre-conscious physical reaction against that which is "improper/unclean" (2). In this sense, it is already linked to the matrix slot of Otherness, to the sphere of non-homogeneity of the Earth rather than the high reason of the Heavens. Most pronouncedly, subjects experience the abject when confronted with the fleshliness of the individual body, as when we find ourselves in the presence of a corpse. For Kristeva, the corpse, "seen without God and outside of science, is the utmost of abjection. It is death infecting life" (4). In the symbolic and actual dichotomy of life/death, death is existentially constitutive of life: "refuse and corpses *show me* what I permanently thrust aside in order to live. These body fluids, this defilement, this shit are what life withstands, hardly and with difficulty, on the part of death. There I am at the border of my condition as a living being" (3). If a human cadaver (from Latin "cadere:" to fall) is the ur-abject, we can see how it features and functions in the theocentric descriptive statement of the Middle Ages in the dichotomy of Spirit/Fallen Flesh, filling the matrix slot of Otherness with the non-believing sinner – metaphorically the original sin, the Adamic *Fall* from the grace of God. In the theocentric terms of Christianity, however, abjection still encounters "a dialectic elaboration, as it becomes integrated in the Christian Word as a threatening otherness – but always nameable, always totalizable" as sin (17). Transuming the descriptive statement of what it means to be human into modernity, the totalizable Sin, the Flesh fallen to the "nonhomogenous nadir of the earth" (Wynter, "Unsettling" 274), the matrix slot of (absolute and non-Human, non-living) Otherness becomes the sphere of Blackness, which we must understand as a praxis: as whatever white modern subjects *do*. While we can maintain the dead fleshliness of the corpse as Kristeva's ur-abject, we must also acknowledge that actual and symbolic death as well as the human-as-flesh has been cast (off) by the racializing theater[9] of modernity as Blackness in order to create its

---

9 Applying this grammar and lexicality, Lyotard's notion of the theatrical volume comes to mind, which Ashley Woodward describes as follows:
> Lyotard explains the relation between libido-desire and wish-desire with the figures of the libidinal band and the theatrical volume. In itself, libido operates according to the primary processes (the unconscious), which knows nothing of negation, space or time. This is envisaged as the libidinal band, a Möbius strip (without inside or outside), along which intensities run at infinite speed, such that their position cannot be localized in time or space. Wish-desire is described according to the slowing, folding, and hollowing out of the libidinal band to form a "theatrical volume." This describes wish-desire as a function of representation: that which is lacked is represented on the stage, while the real thing lies outside the theatre walls. The theatre also describes the secondary psychical processes:

subjects qua sociogeny. From here on, we can no longer merely hypothetically (and pointlessly) conceive of abjection beyond Blackness, but must read it today (corresponding to Wynter's notion of a "truth-for") in its modern, culturally specific actualization as an *abject-for* white subjectivity. The abjective act of modern white subjecthood must be framed by way of its culturally specific actualization as anti-Black by definition.

Importantly, the abjective act is the specific point at which the subject enters language and an "I" can be experienced, uttered, come to exist. In a psychoanalytic framework, this corresponds to the moment of severing the child from the nurturing breast of the mother and the introduction of a symbolic system: "There is language instead of the good breast" (Kristeva 45). By way of abjection, we create the first differential split into me/not-me as the ur-sensation of entering the sphere of this/not-this, of signification at large. It follows that whatever is discursively and affectively mapped onto the experience of not-me denotes the sphere of unintelligibility, which is both the ground for and a threat to signification in the actual, historically contingent, culturally specific subject-making system through which it is articulated. Phobic objects emerging from the abject sphere are thus prefigured (and cathected) by the cultural specificity of what it means to be human, and will always stand for a "hallucination of nothing," turning into a "metaphor that is the anaphora of nothing" and might pose a threat to the discursive system (42). Accordingly, with sociogeny in mind, we can understand how the continual praxis of anti-Black abjection is crucial for the maintenance of modern subjectivity. Given that the (active) psychic, visceral, and cognitive experience of abjection establishes the first not-me and thus creates the *experience of I*, and that "I abject *myself* within the same motion through which 'I' claim to establish *myself*" (3), the modern West is hopelessly dependent on the continual abjection of Blackness to maintain what has originally been established as *I*, that is, as the subject of its social order. Anti-Black abjection, then, lies at the border of symbolic modern being in the world. White subjectivity relies on the continual, structurally prescribed praxis of anti-Black abjection to generate, organize, and perform social signification *as such*. Without anti-Black abjection, the modern subject would be like Kristeva's "child, who has swallowed up his parents too soon, who frightens himself on that account, 'all by himself,'" whose fear "permeates all words of the language with non-existence" (6). The powers of horror become the powers of Blackness. Kristeva considers the "limit that turns the

---

space, time, the concept of language, all of which depend on the basic distinction (this/not this) absent from the libidinal band. (128)

speaking being into a separate being who utters only by separating – from within the discreteness of the phonemic chain up to and including logical and ideological constructs." She then asks: "How does such a limit become established without changing into a prison" (46)? This prison, or "fortified castle" (47), describes a hypothetical state – real only for the borderline patient – in which the subject/object relation has lost all connection to the abject and consequently to itself, having originated in abjection. There is only pure signification, pure structure, "pure and simple splitting, an abyss without any possible means of conveyance between the two edges" (47). Kristeva casts abjection as preventing such "petrification" because it lets "current flow into such a 'fortified castle.'" As a result, a "subject-effect – fleeting, fragile, but authentic – allows itself to be heard in the advent of that interspace, which is abjection" (48). Such a subject-effect is driven by jouissance, which lies in experiencing oneself as both prior to and having *mastered* the initial abjection of separation from the mother.

Kristeva differentiates a generalized and "ostensibly supracultural" (Wynter, "Unsettling" 288) human existence from beasts by way of our capacity to abject, as the initiation into the symbolic (and exclusively Human) realm: "The abject confronts us [...] with those fragile states where man strays on the territories of the *animal*" (12). There is quite a lot to unpack in this assertion. First, we need to observe (meta-theoretically) that Kristeva draws on the distinction between the rational Human being and the "territory of the *animal*" as a pre-civilized state of nature. This very distinction can function only within the dichotomy that creates Man2, the biocentric Darwinian conception of being human, itself already racialized. If, for the abjector, the territory of the animal is a "fragile" state, this means that the abjector distinguishes themselves from a "natural" state of being they (allegedly) once inhabited. The racial ascription is, of course, that the Black body *still* inhabits this space, which means that abjection is a praxis for which a Black-racialized body has no capacity. Kristeva's universalized approach has a colossal Black spot. Moreover, this "fragile" state of abjection, walking the limits of Humanity as it were, speaks to an existential, narcissistic, and solipsist crisis that leads nonetheless to the jouissance of the "subject-effect:"

> [The abject] takes the ego back to its source on the abominable limits from which, in order to be, the ego has broken away – it assigns it a source in the non-ego, drive and death. Abjection is resurrection that has gone through death (of the ego). It is an alchemy that transforms the death drive into a start of life, of new significance. (Kristeva 15)

Relating the joyful subject-effect of abjection as an "alchemical" procedure to culturally specific anti-Black abjection, we can make sense of innumerable white actions otherwise difficult to comprehend (and I consider several of them later). For Kristeva, the jouissance of the "subject-effect" transcends superficial concepts of fear or desire: "jouissance alone causes the abject to exist as such. One does not know it, one does not desire it, one joys in it" (9). This is in part why it is so difficult to grasp: the subject may experience jouissance in the perception of Blackness or in the performance of anti-Black abjection – voluntarily or involuntarily – without recognizing the racial structure that underlies it. However, in actuality, the anti-Black subject-effect in white modern subject-bodies is achieved through the modalities of fear and want, fear being an "abortive metaphor of want" (35). Fear, in the first instance, denotes the "states of distress that are evoked for us [adults] by the child who makes himself heard but is incapable of making himself understood." It denotes a physical state right in the process of subject making, the "upsetting of a bio-drive balance" that is later articulated in symbolic terms through and repeated by the "constitution of object relation" (33). Being in the world as a subject can thus be described as "alternating with optimal but precarious states of balance" (34). Abjection ensures the maintenance or reinstallation of libidinal (and material) homeostasis in the Human organism. This "calls attention to a *drive economy in want of an object* – that conglomerate of fear, deprivation and nameless frustration" (35).[10] Kristeva states that "[p]hobia literally stages the instability of object relation" (43). For the white modern subject, the foundational phobic or phobogenic object is the Black-racialized body (see also Hartman, *Scenes* 57; Fanon, *Skin* 82–108). Secondarily, this also accounts for other signs that signify Blackness. The most efficient way to actualize the white self and generate a subject-effect authentically (if fleetingly) is by sliding through the portal of abjective experience and delving into the sphere of Blackness. When Kristeva asks how the limits of relationality, of language and the subject/object splitting, can be prevented from becoming a

---

10 It must be noted that, though bearing "the mark of frailty of the subject's signifying system" (Kristeva 35), the phobic object does not come into being by way of substituting the father as the first order object of fear, and does not itself inhabit the sphere of relationality. For both Broeck and Kristeva, abjection provides the ground for object relations but has no access to the very space it enables. Speaking with Wilderson in the culturally specific terms of anti-Black modernity, the dehumanizing and abjective praxis of Black enslavement provides the only "access to (or, more correctly, banishment from) ontology" (*Red* 18). Broeck writes:
> To come into being, the European subject needed its underside, as it were: the crucially integral but invisible part of the human has been his/her abject, created in the European mind by way of racialized thingification: the African enslaved, an un-humaned species tied by property rights to the emerging subject so tightly that they could – structurally speaking – never occupy the position of the dialectical Hegelian object as other, has thus remained therefore outside the dynamics of the human. ("Legacies" 118)

prison, how the ego can experience itself as free, the culturally specific answer is: qua anti-Black abjection.

Where the religious codes for the signification of the abject within the structures of the theocentric super ego provide the concepts "defilement, taboo, or sin" (Kristeva 48), the modern matrix of being Human holds any signifier that denotes "Blackness" as the invigorating source for its white subject's self-making. Blackness is the current that flows in the object relation of white supremacy whenever needed, animating, resuscitating, and invigorating the system of white supremacy on all levels, in all groups and individuals to the extent that their libidinal economy functions according to (and is rewarded by) that system. White modern subjectivity reenacts and stabilizes white abjectorship as the primary mode of its specific sociogenic principle. It is qua specifically anti-Black abjection that the modern subject reenacts what Lacan calls the "unitary bent." By way of anti-Black abjection, we white subjects confront and remind ourselves of a maternal entity, even relive it (Kristeva 13). In our current system of existence, Blackness and all units and modes to which that signifier is attached functions as a sort of portal to the joyful experience of a subject-effect, whether triggered by fear or desire. The want for one's subjectivity overrides either. It cannot be stressed enough that this portal of Blackness is not a natural given but one that the culturally specific subject of white modernity (exclusively!) can open up any time and place. In the bios/mythoi hybridity of our existence, Blackness is located in the "languaging" sphere of mythoi (Wynter, "Catastrophe" 32), which structures, shapes, and informs our bios. Below, I aim to help develop a language for the specific procedures and practices of shaping or informing the human body as an affective biochemical configuration. First, however, I consider some of the dimensions in which anti-Black abjection happens.

Modes of abjection

I have suggested that anti-Black abjective practice, *by whites for whites*, is always already structurally directed against those racialized as Black. It can be an *internal procedure* or an *externalized outward act* – or both at the same time. Further, on the level of the anti-Black libidinal structure ingrained in politics, bodies, embodied psyches, and the sociopolitical sphere, I propose that we need not distinguish specific kinds of anti-Black abjection. Is the willful seeking out and consumption of a given film (and the associated support of the creation and production of such images) not also an act of abjection? Is the creation of a mental image (be it based on memory/imagination or actual perception) not an act in itself, especially given that embodied cognition teaches us that memories, perceptions, and imaginations function similarly in our bodies? Is a feeling not also an outward performance, for example in the microaggressions we as white subject-bodies send out as involuntary but perceptible facial expressions, sweat, or change of heart rate too minute to be easily detected? From what cranny in the body does a white subject derive pleasure in listening to Prince? What exactly does an audience enjoy in the *Black Panther* movie? Is it at all possible for a white audience to decode the affect of such a film in a meaningful way? What does the white collective *audience-we* do with, about, and to Spike Lee's *Bamboozled?* Accepting the polyvalence of abjection (as articulated in this series of questions) is consequential. Certainly the cop with his finger on the trigger at the sight of a Black man is not *exactly* like the auctioneer on the coffle, who is not *exactly* like an audience member of a minstrel show, who is not *exactly* like a blackface performer in that same show, but they are all part of the same sociogenic species, sharing the same discursive and affective DNA. Even though it may be a long way from a semi-ironic white fist bump to the Charleston church shooting, it is important to understand that there is an unbroken continuum, a shared axiomatic and affective ground of violence that connects the two.

Language is the plane on which we can analyze this continuum out in the open. The use of anti-Black abjective language finds its contemporary discursive nexus in the question of whether the use of the n-word by white people is or should be permitted, and if so, in what context(s) it might be permissible. The verve with which white people use (and rationally defend) their right or structural capacity to use the word – attained through the gratuitous and dehumanizing violence of modern Man – can only be understood in terms of psychic necessity, in the symbolic value that uttering and hearing, writing and reading it provides. As Lyotard writes in *Libidinal Economy*, there are some words that the mouth "prefers to utter" (2). The n-word is such a word for whites because its articulation (real or imagined) is in itself an act of anti-Black abjection that provides white subjects the joy of a subject-effect. It is one of the most effective pathways of modern sociogeny because it encapsulates white subjectivity in the entirety of its historical legacy. In the line of argument presented here, the n-word is the ultimate pathway for reasserting not only white superiority, but one's white existential subjectivity and subjective existence. Abjection is in action every time a white subject speaks, hears, imagines, writes, discusses, debates, defends, or even critiques the word.

Tellingly, the debate about using the n-word in improv (or the comedy industry at large) hardly ever addresses the question of its affective meaning and effects as considered in this project, once again using Blackness to ponder political notions of freedom of speech and artistic (or satirical) license. These feature largely in the discursive self-description of improv, but one central argument is often put forward, namely the idea of "playing true to character," even if that *spontaneously discovered* character is racist:

> *Perkins:* People often use that to justify themselves. They think that because it's comedy, it gives you leeway to say whatever you want. But at the end of the day, we're two humans on stage making things up. You don't forget that you're a person. You can play a character, but that's what it is. It's a character. It's not you. You know what can and won't hurt your partner. Don't cross that line. I don't think that hurting another person is less important than the truth of a character. Once the scene is over, we have to live as people. First and foremost, when I am playing, I am thinking that nothing is ever that serious where I am not in a space with another person.
>
> *Bullock:* You don't have to use that word to portray a racist character. I feel like the idea to "play the truth of my character" is a cop-out. Words mean things. Words have meanings. Once you say it, the whole scene has to be about that word. That's the kind of word it is. It has that kind of weight. So if you're bringing it into a scene, you better be prepared for that scene to be about that until the scene is done. If that's your point of view, that you can use that word to play "true to character," or even just to get a reaction, if you use it without foresight of what that word means, if you can't handle the context of that word or understand how it works in a room, then I'll be over here not doing that. If I hear that you use "playing true to your character" is your excuse, then I will blacklist you in my mind.
>
> *Perkins:* Oh yeah. Definitely. (personal conversation)

> *Schleelein:* I was discussing the matter of the n-word on stage with Rob Wilson last night.
>
> *Arashiba:* Is Rob black?
>
> *Schleelein:* Yeah. And the argument it came down to was: Is context important? Can a white person say that word without any repercussion? What is the exact context, if that is okay?
>
> *Johnson*: Me personally, I say no.
>
> *Schleelein:* Of course. It'd have to be really funny, right? Maybe not. But this is the thing we were talking about: context. If you can't understand the idea that at some point it's funny and that at some points it's not, and you don't appreciably recognize the difference, then fuck you. You're out.
>
> *Question:* What kind of context would make it funny, in your opinion?
>
> *Arashiba*: God knows.
>
> *Schleelein*: I don't know. Dave Chapelle could probably do a good one.
>
> *Johnson*: Context for a white guy saying it.
>
> *Schleelein*: Oh no, for white man – that'll be such a fine line.
>
> *Johnson*: Repeating a black guy saying it. Right there, as the black guy said it.
>
> *Schleelein*: Yeah. There'd have to be a level of protection. If you're at all a decent person to make that joke. (personal conversation)

This debate also relates to the question of what white people are qualified to talk about, especially if it involves scenes representing enslavement:

> *Arashiba:* It's higher mathematics. You know that in some ideal world, a white guy can make a slavery joke.

Johnson: Sure!

Arashiba: What's the math? What's the setup? How do you get to that so that it's funny and it's true, and it's not a stupid caricature?

Johnson: As long as the person doing their bit is being genuine about what their story is. Like Derek, this white guy here, he's doing a really great white Muslim. His white Muslim character is literally grounded in the reality of what black progression is. And that matters. It matters if you're grounded in the reality of what your situation is. Say you're a pony. If you're playing a pony on stage, but you care about pony shit, you care about your pony, you can talk about that pony's feelings eating gummy bears. I hope that is not too weird. You have to care about what you're playing. At least myself, when I am doing my little improv shit – as long as I am invested in whatever it is, it'll work, and it'll be funny. (personal conversation)

I believe the term's performative and affective function is lost in this debate on the moral (il)legitimacy of white people using the n-word. Those referred to and affected by the use and denotation of the n-word are not – discursively – part of that political sphere, so their position cannot be made entirely intelligible for those living *as* the white referent-we for which anti-Black abjection is *constitutive*, as in the use of the n-word. Letting go of the capacity to apply, imagine, and perform this term, so instrumental in the generation of white modern subjectivity, would involve the disintegration of our current subjectivity and thus challenge Western sociability at large. When sociopolitical subjectivity and psychic-as-physical integrity is at stake, the culturally specific bios/mythoi praxis of contemporary Western being Human as Man must industriously seek lines of defense for using this term. By using the n-word – and all other anti-Black procedures – we white people undergo an abjective experience with the least possible effort. And more often than not, we are not even aware of it. We can therefore think of anti-Black abjection as action, perception, and imagination. I suggest that even the act of imagining the n-word functions as anti-Black abjection; while no outward act violates Black-racialized bodies, it still activates the substantial (and perhaps more efficient) dynamic of the modern sociogenic principle as it pertains to modern Man's biopsyche. This I believe to be true for all elements or modes, artifacts, or gestures racialized and marked as Black by the white subject-body. The way in which the white subject draws upon, fears, partakes in, or desires Blackness for its own self-making, aiming at the subject-effect qua abjection, can occur through conscious acts or involuntary abjective sensations.

Affective transumption

A telling and relevant incident took place at an improv workshop I attended in Bremen in 2014, when two white German improvisers used the term *digger / digga / diggah* in a scene. The term can be spelled in different ways and entered the German language via the Hamburg hip-hop scene in the 1990s. While I never write out the n-word, I do write out this transumption of it. Nobody would understand what the "d-word" means, which is precisely the point I make in this section. In (semi-)journalistic considerations of the term *digger* or its variants, it is commonly argued that it derives from the German *Dicker* (meaning "thick one"), a functional appellation for people with whom one has a close, trusting, *thick* friendship. Phonetically, this is plausible, as the Hamburg accent softens hard consonants in the middle of a word. Some attempts to historicize the usage of the term invoke the so-called Barmbek Basch dialect in specific working-class milieus in Hamburg in the early twentieth century. Today, the term is a "perennial favorite in teenage slang" (Assmann, my translation), and it is regularly suggested that "unlike many other rap terms it has not been imported from the USA" because the English term digger has a different meaning (Kurby, my translation). These arguments, however, miss the point. Even if the term is of German origin and cannot be traced back to the English word etymologically, the logic, validity, and significance of this position must be called into question. The d-word has gained momentum since its early German hip-hop usage, and because the predominantly white connoisseurs of predominantly white early Hamburg hip-hop culture have now grown up, it is now widely used throughout Germany in all sorts of milieus, sometimes more and sometimes less ironically. (Irony is, of course, insignificant with regard to the libidinal effect of a term's usage, but is often used to justify the specific way of attaining the very same biochemical effect, not unlike the fist bump so commonly used these days with varying degrees of irony but with great performative stability.) In the workshop, two male players improvised a German-language scene using that word repeatedly, presumably poking fun at what they presented as a deranged hip-hop youth culture – with a playful regional feud between Bremen and Hamburg also feeding into it. When the scene was done, two white Americans in the workshop asked what that was about and were met with silence. They were confused by what was, for them, a shamelessly racist performance of stereotypically Black demeanor and the repeated use of what they understood to be the n-word. The Germans in the room were quick to explain the etymological and regional arguments to defend the term's use, and it was left at that.

The arguments they put forward may or may be etymologically correct. Yet given the contemporary use and function of the term, they are of no relevance.

The conceptual and functional overlaps between the d- and the n-word are too obvious, as are the hip-hop context and its specific aesthetics of expression, the performative gestures and adopted attitudes that accompany it. The two common ways of spelling it are too similar (with *-er* or *-a* at the end), and the term is too immediately a specific expressive gesture within an already anti-Black cultural context of Black imitation/obliteration. The d-word cannot reasonably be delinked from the n-word. (To do it all the same requires active white abjective ignorance, which is what we observe here.) Importantly, as this example shows, the term is used predominantly among white people in contexts and situations in which they imitate what they subliminally understand as Blackness. It is an anti-Black abjective term that thrives in an anti-Black abjective cultural and in the highly specific framework of white German hip-hop culture. Like Blackness, it is open and fungible, available for anyone. At the same time, it can function as a means to aggrandize Hamburg hip-hop culture (and its representatives), supporting claims for that city as the originary and the most authentic local hip-hop culture in Germany, the presumed original gangstas so to speak, against its widespread use. In one of its more recent and widespread articulations, the song "Ahnma" by Beginner makes this claim: "Everyone says 'Dicker' nowadays / We're putting Hamburg back on the map" ("Jeder sagt Dicker heutzutage / Wir packen Hamburg wieder auf die Karte.") Though this example might appear to be locally specific and structurally insignificant, it points to something larger: the transumption of an affect trigger to the point of timeless unrecognizability. That is, the transumptive process doesn't change the affective function, but adds more and more layers to the veil that hides its anti-Black origin. Only a small selection of the speakers who use the term would consciously think of the n-word, which many would even consciously avoid for its racism. However, the effect that the d-word has on a white subject's body is similar if not the same: a close relative or a derivative of it, perhaps slightly less intense. Etymological arguments can never account for the power and popularity of the term today. Reading it within anti-Black abjection can. What we can take from this brief consideration is the idea that *abjective transumption* can account for anti-Black abjection without Blackness being *visibly mobilized*. A discursive entity with abjective powers for whites appears to be entirely stripped of its originary Blackness. This is a transumption of anti-Black affect into non-recognizability of the n-word as its ur-trigger.[11]

---

[11] To be clear: I am not making any statement about the legitimacy of using the term, or whether this legitimacy could or should be raced in the sense that white people or Black people respectively should or should not use it. This is merely a consideration of the structures in which this term exists, and which ensure its discursive fortitude. On the same note, in two in-

3.3 Embodying anti-Black abjection

Sociogenic grounds

Neuroscience is on the march and has begun to do important work, proving the existence of what is regularly called "implicit racial bias." Studies consistently show that unconscious racial bias has real-world effects, for example in police shootings (Hehman et al.), employment (Pager et al.), healthcare (Hoffman et al.), or infant health (Orchard and Price). Such studies are significant for raising awareness. However, they all *implicitly* draw on the white default position, because unconscious racial bias is only implicit for those whose white universalized bodies perform it. Those affected by it negatively do not need such scientific proof. Consequently, these studies, on the whole, are slowly exhausting themselves in the *discovery* of more and more fields in which unconscious racial bias can be observed. Further, the rhetorics of generalized bias, of which the racial variant is one among others, quickly fall into the rhetorics of relational discrimination, which do not communicate with an Afro-pessimist theoretical framework. However, in her theorizations of the sociogenic principle, Sylvia Wynter regularly draws on the findings and perspectives of cognitive science and substantializes her theorization through neuroscientific empiricism. There are good reasons for this: the whole field I address as embodied cognition seems to stand right where Wynter explores the ways and workings of sociogeny. Embodied cognition can indeed be mobilized to analyze what happens in the interplay between bios and mythoi, helping us to concretize the ways in which language, words, and stories shape and inform our biological existence, which in turn prefigures how we can think and feel about the world we live in. In this section, I briefly introduce a few central ideas behind embodied cognition and the trope of the *body-brain* to denote a body that corresponds to Wynter's idea of being human in bios/mythoi hybridity. With recourse to the influential neuroscientist Antonio Damasio, I zoom in on the concept of somatic markers. Referencing the discussion above, I hypothesize that we can conceive of anti-Black

---

stances in this project I had used the racial terms "dreadlocks" and "dreading hair." Even though I was fully aware of the meaning of the verb "to dread" and the adjective "dreadful," in this particular context I always understood it simply to mean "twisted hair." There is no German equivalent of the term, so for me it seemed a neutral loan word for a hairstyle racialized as Black. In other words, my white epistemological position *enabled* me simultaneously to know and not to know the connotations of the term – and *empowered* me to ignore the fact that there are no neutral terms in a racialized domain. I am grateful to my editor Annie Moore for pointing this out to me.

abjection as the continual praxis of generating psychobiological homeostasis, that is, the somatic state of a subject-effect or a biochemical "sense of self," to use Wynter's phrase ("Principle" 54).

As the nomenclature suggests, embodied cognition helps us think of the rational and the biological functioning as a "single unit," against their differentiation in Cartesian tradition (Claxton 78). Theorists of embodied cognition – behavioral neuroscientists, neurologists, biopsychologists – provide evidence for the fact that "the body is self-governing" with no "big boss in the brain who forces through resolutions and dictates policy" (80). The brain does coordinate a lot of information: electrical impulses travel along a looping trajectory via the "autonomic and central nervous systems to and from the brain [...] chemical messengers that flow through blood stream and lymph system," including physical information about body movement and the wider as well as the microscopic scale (81–82). The entirety of this "maelstrom of physico-electro-chemical activity" comes together in the brain (87), where the various systems and their "ambassadors" confer (89). Yet the brain is one bodily function or system among others. While consciousness may be said to reside in the brain, this organ is no less physical, biological, or chemical than the rest of the body. In reference to the transumed dichotomy of High Reason/Irrationality, then, we must assert that the higher faculty is not so high after all. Further, and in keeping with a long tradition in affect theory, for theorizers of embodied cognition the body is not fixed and stable but in constant motion.[12] Claxton states that "we exist by happening" (36), and that "the human body is a verb" (54). For him, this means that our information processing procedures are constantly present and continually updating, even though certain information can be foregrounded over other information within the brain. He applies the metaphor of a wave:

> On the gross level, the wave keeps its form, but the water – the content – keeps changing. At any "moment," the wave represents the sum of total of all currents, swells and winds that are acting at that location. They come together to create a particular wave-form, with its signature composition and direction. Waves have a width; that is, they integrate the forces acting not at a point but over a small region of the ocean. The biological constraints of this span of integration might well account for the tenth-of-a-second duration of these apparent "moments." But each momentary wave is not separated from the moment before and the moment after. Like a real wave, it has both a leading and a trailing edge. It is simultaneously rising, existing and fading. In the rising are expectations and predictions of

---

[12] At least since Spinoza, the forefather of affect theory, the idea that living bodies are not fixed and stable but in motion has been developed in many theoretical avenues. For Wynter, too, the destabilization of Humanism by de-conceiving it as a praxis of being human is central. This position also provides the axiomatic ground for the theorization of abjection as praxis.

what the future may bring, and in the fading are the echoes of the confirmations and surprises that arose from those moments that have just been. [...] (We might, if we're feeling whimsical, see the properties of the seawater itself as the capabilities of the body to behave; the current and groundswells as the values and concerns in play at that location in time and space; and the winds as the influences from the external world). (91)

The body-brain decides from moment to moment "what is the next best thing to do" (10). How and why human beings decide what to do in a given situation, what drives their behavior and structures their way of being and behaving in the world, is central to the study of embodied cognition. Together with the awareness of the body-brain functioning as a single unit, we are engaging with how "the 'decision-making' of the brain can be influenced by a badly behaved bacterium in the gut, and the level of sugar in the blood can be altered by a squeak or a dream" (4–5). Without going into much more detail, let me emphasize with Claxton that "the brain is not just in the business of telling the body what to do. [...] Body and brain are tied together so intricately and so rapidly that it makes no sense to locate all the 'intelligence' in one and none in the other" (87). It must be understood that "intelligence can be embodied in physical structures, and that some structures can therefore take some of the strains of minds and brains" (41). This, in effect, breaks down the modern dichotomy (and its theocentric and ancient predecessors) that prefigures the white subject on the grounds of Reason as opposed to the sphere of Blackness and irrationality. Moreover, "much of our somatic intelligence operates unconsciously, without conscious supervision or even awareness" (Claxton 7). Though perhaps not entirely zombies, robots, or computers, we are certainly far more limited than many are willing to accept or even realize.[13]

Importantly, embodied cognition can tell us a lot about the function and functioning of emotions. No longer are they understood as separate from rationality. Emotions "involve muscles and glands, blood, sweat and tears, as well as thoughts, memories and imaginings" (Claxton 103). They are the "most obvious place where we experience our bodies, brains and minds coming together." Rather than separating emotions from rationality, "[w]e must begin by seeing emotions as contributing to our ability to act intelligently, not as impediments to such action" (103). Claxton considers emotions "deep, bodily-based constituents of every kind of human intelligence" (104). For proponents of embodied cognition theory, emotions are "built-in 'default' settings of our whole embodied systems" (104), automatically activated as responses to significant events:

---

[13] Note, though, that the recognition of these limitations decidedly does not imply their transcendence. The author of these lines also has a body.

> If we see something as an object of desire we are automatically primed to approach and secure it. If we are hungry we start looking around for sources of food. [...] if we see an aspect of our world as dangerous, we prepare to avoid or neutralise it. [...] Though they come in many shades, emotions are intelligent responses to events that are relevant to what we value – and what we value has its roots firmly in the physical body. (105)

An emotional state is a response to a cue that signifies such important events. They are not "not actions in themselves, but anticipatory states of readiness to respond to events that, we suspect, might be about to unfold." These states of readiness are pre-settings of affective configurations, not unlike those "that come with the audio amplifier on a modern TV" (106). They combine biological, chemical, and physical phenomena that manifest in and through the body.[14] Yet the recognition of these aspects, and presumably many others not mentioned here, should not leave the impression that they are hierarchically structured or that there is a certain protocol in their sequencing. Embodied cognition theory recognizes that – even though different parts fulfill different functions – the whole body is part of this configuration, and it is unhelpful to seek to disentangle what the brain or other parts of the body do on their own:

> These whole "body+brain+sensors" reactions are so intricately interwoven that it is impossible for us to pull them apart and tell what is "cause" and what is "effect." The circular loops connecting body and brain are bi-directional, so that "higher" processes are influencing "lower" ones, at the same time as the "lower" are feeding information up to the "higher." Words like "resonance" and "reverberation" capture this shimmering complexity much

---

[14] To get a clearer idea of what this means, I quote the paragraph from Claxton in full:
Which bits of our bodies do emotions engage? The broad answer is: almost all of them. But it may help to highlight some of the most important. First, our internal physiology can be altered. [T]his can include changes to heart rate and blood pressure, rate and depth of breathing, the physical and chemical behavior of the intestines, and the chemical composition of the blood and lymph. In addition, aspects of bodily posture, facial expression and voice quality can change. Our shoulders may drop, or faces become angular and forbidding or tender and loving, and our voices grow hard or soft, loud or quiet. The skin changes color as its blood supply increases or decreases – our cheeks burn with embarrassment or become pale with rage – as well as changing in sensitivity and sweetness, and our body hair can stand on end. The big action muscles of arms, legs, shoulders, neck and so on can be tensed or relaxed, and fingers get ready to ball into fists or stretch out into soft instruments of caress. Actions may become slow and ponderous, or sprightly and vivacious. Sensory muscles are affected: eyes move in their sockets, pupils dilate or contract, nostrils twitch [...] Eye contact is penetrating or steely, or the gaze averted or shy. And the brain sets up patterns of expectation and prediction: some constellations of attention, memory, thought and imagination become primed; others may fade into the background. (107)

better than ides of "stimulus" and "response." Did I see the bear, feel afraid, and so tell my legs to run? Or did the unfolding of bear-seeing, gut-trembling and leg-thrusting happen in such a fast and loopy fashion that they are, in fact, different facets of the same holistic episode? Traditional "folk psychology" tells us the former. Embodied science tells us the latter. (106)

The body experiences various emotional modes that involve the body-brain – Claxton lists distress, recovery, disgust, fear, anger, sorrow, shame, desire, inquiry, care, and anxiety, though others suggest other groupings – and these are variable in themselves. (In the pre-set of *Disco,* you can still add bass.) As a consequence, there are infinite combinations and configurations for refining, nuancing, and modifying an emotion. Conceptualized as biophysical *modes,* then, emotions "can operate perfectly well at the physiological and behavioral levels without any need for conscious supervision, or even awareness" (118). Regardless of how individually refined or sophisticated emotions are experienced, and whether or not they reach consciousness, "they retain their rootedness in the physical workings of the body" (121). Human decision-making – figuring out the next best thing to do – must be viewed in light of this physiological reality. Not only is so-called "rational thinking" fundamentally physical (that is, chemical and biological), but the body very often knows more and faster than logical processing could function on its own – if it could function without it at all. Indeed, the body often knows better and learns faster without the mind, even in matters traditionally related to the mind. Even though decisions are ultimately made in the brain, they are not rational. The brain itself is more like a conductor of a heterogenous choir and decides which songs to sing at the concert: "Each of the possible courses of action under consideration gets tagged with a number of indicators that will help the brain to do its job of conducting the somatic orchestra" (Claxton 98).

Somatic marking of Blackness

In *Descartes' Error,* Antonio Damasio conceptualizes and empirically substantializes emotions as *visceral* – a tremendous contribution to the study of embodied cognition. What Damasio contributes to this project is his differentiation between what is natural, what is learned, and what is cultured. He also distinguishes between primary (innate, "early") and secondary (learned, "adult") emotions. Primary emotions are "pre-organized mechanisms" (a term borrowed from William James) wired into our bodies *before* we learn anything. Without the interference of consciousness, our bodies react to stimuli from within or outside themselves.

These are not necessarily concrete objects but rather features such as size (large animals), a particular type of motion (for instance, reptilian), or sounds (like growling). Such stimuli – or combinations thereof – can be "processed and then detected by a component of the brain's limbic system, say, the amygdala; its neuron nuclei possess a dispositional representation [of those stimuli] which triggers the enactment of a body state characteristic of the emotion of fear" (131). The whole procedure may remain outside consciousness to the degree that one does not need much knowledge about the object, or even to recognize it; "all that is required is that early sensory cortices detect and categorize the key feature or features of a given entity" (131). An emotion is thus best understood as "a change in your body state defined by several modifications in different body regions" (135). Damasio's explication clarifies how this conceptual framework resonates very strongly with the libidinal economy and its homeostatic ideal:

> As a whole, the set of alterations in the body defines a profile of departures from a range of average states corresponding to functional balance, or homeostasis, within which the organism's economy operates probably at best, with lesser energy expenditure and simpler and faster adjustments. (135)

With Damasio, I understand an emotion "as the collection of changes in the body state that are induced in myriad organs by nerve cell terminals, under the control of a dedicated brain system, which is responding to the content of thoughts relative to a particular event or entity" (139). Some of these changes are perceptible by an outside observer (such as body posture, ways of moving, or sweat intensity), while others are "perceptible only to the owner of the body in which they take place" (139). All of these serve the homeostatic ideal, material, psychic, libidinal. How do we get from primary, early childhood emotions to secondary, adult emotions?

For a secondary emotion to be realized, we need to be able to feel it. The emotion, a specific change in the configuration of the continual affective wave, must reach consciousness. Once we become conscious of what happens in the body, we can subtract further information from the emotional event in relation to what triggered it, which helps us to apply that knowledge in the future to make different decisions. Damasio writes:

> Feeling your emotional states, which is to say being conscious of emotions, offers you *flexibility of response based on the particular history of your interactions with the environment.* Although you need innate devices to start the ball of knowledge rolling, feelings offer you something extra. (133)

Where primary emotions serve as the basic structure of human emotional life and our personal and social behavior, they "are followed by secondary emotions, which occur once we begin experiencing feelings and forming systemic connections between categories of objects and situations, on the one hand, and primary emotions, on the other" (134). Whereas the limbic system and the amygdala suffice for the primary emotions to function effectively, other parts of the brain capable of representation are applied for secondary emotions: "the network must be broadened, and it requires the agency of the prefrontal and of somatosensory cortices" (134). What comprises an emotional experience of the second type? First comes a "conscious deliberate consideration you entertain about a person or a situation" (136). These are representations of images, some verbal, others non-verbal. Then, non-consciously, "networks in the prefrontal cortex *automatically and involuntarily* respond to signals arising from the processing of the above images" (136). Because these are reactions to a representation, they are necessarily *acquired* rather than *innate*, "although [...] the acquired dispositions are obtained under the influence of dispositions that are innate" (136). Then, "non-consciously, automatically and involuntarily, the response of [these] prefrontal dispositional representations is signaled to the amygdala and the anterior cingulate" (137), which causes the body to react: "viscera are placed in the state most commonly associated with the type of triggering situation," and the motor system and skeletal muscles express that emotion by completing the "external picture" (138). Chemicals are released that result in changes in the body-brain. An actual situation can trigger them by way of memory or imagination. The representational faculties within the brain are the same ones that trigger the biochemical reactions making up the emotion.

The emotion as such, the tidal wave that floods the body, is a different phenomenon from its conscious perception and interpretation, let alone from the linkage between the emotional wave and the signal that caused it in the first place. For this process as well as its outcome, Damasio reserves the term "feelings." For a body-brain to register an emotion consciously, it must be represented in the brain and signaled "through nerve terminals that bring to it impulses from skin, blood vessels, viscera, voluntary muscles, joints, and so on" (143). The somatosensory cortices in the insular and parietal regions "receive an account of what is happening in your body, from moment to moment," and thus "get a 'view' of the ever-changing landscape of your body during an emotion" (144). Parallel to this neural information, information also comes through a different channel, a "chemical trip": "Hormones and peptides released in the body during the emotion can reach the brain via the bloodstream, and penetrate the brain actively" (144). This means that the brain gets information not only about what is happening in the body, but is also *instructed* about the way it should work with or respond to that information: "What gives the body landscape its character at a given moment is not just a set of neural signals but also a set of chemical signals that modify the mode in which neural signals are processed" (145). If the essence of emotion is the basic experience of the body, the essence of feeling is the "process of continuous monitoring, that experience of what your body is doing *while* thoughts about specific contents roll by" (145). Feelings are thus what happens when emotions are consciously attached to the signals that cause them: "a feeling depends on the juxtaposition of an image of the body proper [i.e., the current body state] to an image of something else, such as the visual image of a face or the auditory image of a melody" (145). A feeling, then, is always *about* something; it links an emotion to an external stimulus and lets us reflect on that link.

Notably, the image of the signal and the subsequent (!) image of an emotion must be kept separate. They do not become one. And like a sign and its referent, they are arbitrary. The combinations may be "unexpected and sometimes unwelcome. Their psychological motivation may be unapparent or non-existent, the process arising in a psychological neutral physiological change" (146). The separateness of the two representations not only "affirm[s] the relative autonomy of the neural machinery behind the emotions" but also "reminds us of the existence of a vast domain of nonconscious processes, some part of which is amenable to psychological explanation and some part which is not." This is why white people cannot but actively abject Blackness when we encounter it. When Hillary Clinton stated during her 2016 presidential campaign that "Black men can be scary," she addressed and universalized a *culturally specific* white feeling, acknowledging an emotion that (structurally) only white subjects know without recognizing it as a white solipsist event rather than a feature of Blackness, something that Black-racialized bodies are assumed to be. Understanding the difference between emotions and feelings might have instructed her to do otherwise. But even if I can ponder *why* I feel a certain way about a certain thing, this does not mean I can control the emotion that underlies the feeling. I can judge it, be ashamed or proud of it, but I cannot change it voluntarily, even with the best of intentions. In the world as we know it, we will always remain voluntary or involuntary anti-Black abjectors.

Racializing the subject-aeffect

Bringing psychoanalysis in conversation with embodied cognition, I posit Kristeva's notion of subject-effect as an emotion understood in the language of embodied cognition. In order to mark this biochemically affective dimension in its culturally specific relation to abjection, I modify Kristeva's term "subject-effect" slightly into a "subject-*ae*ffect." The ligature is a helpful reminder that we are concerned with a *subsequent consequence of a culturally determined action or perception* as well as its embodiment. Understood as a secondary emotion in Damasio's sense, the subject-aeffect is grounded in a "conscious deliberate consideration you entertain about a person or a situation," such as the perception of a signifier (verbal, sonic, linguistic, physical, imaginary, or real) marked as Black, followed by the non-conscious activities of "networks in the prefrontal cortex [that] automatically and involuntarily respond to signals arising from the processing of the above images" (136). I locate the subject-aeffect in the very space where a secondary emotion draws on the nonconscious workings of the body-brain to *discursively naturalize* an affective wave, even though the linkage between the trigger and the effect is arbitrary. Damasio writes:

> Nature, with its tinkerish knack for economy, did not select independent mechanisms for expressing primary and secondary emotions. It simply allowed secondary emotions to be expressed by the same channel already prepared to convey primary emotions. (139)

Crucially, I believe that this accounts for genuine misinterpretations about the world and ill-fated debates about whether "Black Men can be scary." What *feels right* in the body is not necessarily natural and correct, just because the body is an object of nature. With Wynter, we must understand our subjectivities as well as our body-brains as bios/mythoi hybrids. The necessarily abjective (imagined or real) encounter of a white subject with a thing, body, mode, image, or other phenomenon signifying Blackness is both conscious (it must be *read* as Black) and unconscious because, as whites, we may not deeply understand the emotions it gives us, but remain on the level of feelings, registering a bodily sensation linked to a given perception. We may consciously describe a fear of Black men, observe an emotional-sexual inclination toward racialized porn, or interpret ourselves intellectually as connoisseurs of jazz; we may fear Black-racialized neighborhoods, feel revolutionary copying hairstyles, or imagine our own transcendence when we improvise. In the register of feelings, white subjects can discuss innumerable nuances of the subject-aeffect as a somatic state, but it remains grounded in subject-aeffective emotions structured by anti-Blackness. The visceral experience, the affective wave I name "subject-aeffect," the exclusively white experience of one's own physical and symbolic integrity, the "sense of self" (Wynter, "Principle" 54) is not an action in itself – it is caused by anti-Black abjection, which, as a structurally prescribed (inward or outward) performance, happens on the level of sociogeny. We can understand it as a pedagogical procedure that a) teaches the white subject of modernity "how to be a good man or woman of its kind," (Davis qtd. in Wynter, "Unsettling" 271), that is, how to have a subject-aeffect, the visceral experience of a sense of self, and b), permanently ingrains this knowledge in the white subject's body-brain.

Language is essential to this procedure. Through language, the infant enters the symbolic system and gains access to its affective, emotional repertoire. When Aasia Bullock observes (about the use of the n-word on the improv stage) that "words mean things" (personal conversation), her claim resonates with Claxton's assertion that "[w]ords are ways of activating neural circuitry (and altering biochemical processes) through speech and writing" (147). Any affective occurrence in the body shapes and trains the emotional registry. While there is something to be said for life-long learning, the affective repertoire in which we have been trained (and the semantically arrested cathexis that structures our individual and collective libidinal economy as it develops throughout infancy, childhood, and adolescence) powerfully hardwires all our decision-making and emotional structures. It defines what homeostasis means for white subjects within the white modern culturally specific genre of being human as Man. From the moment we can isolate a word or phrase and attribute an experience (observed, undergone, or both), from our earliest engagement with or perception of the n-word (or any other signifier affectively read as Blackness), our neural circuitry programs us to be the anti-Black abjector-subjects that the matrix of modernity prefigures and demands. It hardwires us biochemically to attain homeostasis easily via anti-Black abjection. I think this is part of what Sylvia Wynter addresses in theorizing the sociogenic principle and how *logos* and *mythoi* concretely inform our being as *bios*.

Somatic solipsisms of the languaging white body

Damasio's somatic marker hypothesis posits that part of our decision-making is motivated by conscious or unconscious knowledge (and thus ignorance) of the possible outcomes of various reactions to objects, situations, events, people, or circumstances. Arguing with Pascal that "[w]e almost never think of the present, and when we do, it is only to see what light it throws on our plans for the future" (qtd. 165), Damasio explains that in any given situation that demands action, our mind is "replete with a diverse repertoire of images" about what could happen if we did this or another thing (170). This corresponds to the moment-to-moment decision-making Claxton describes as the continual assessment of "the next best thing to do" (10). Damasio thinks of it as the "selection of a response option" and similarly departs from the Cartesian or Platonic "high-reason view" of decision-making, that is, the idea that Human beings as rational subjects can arrive at ideal assessments of situations purely by intellectual labor detached from sentiment (165). He emphasizes that decision-making is largely based on unconscious and somatic preselection. His argument primarily runs along the lines of reducing response options on the grounds of negative experiences. If a certain response option to a given stimulus has been experienced as sufficiently negative, that option will automatically be dysselected before we have the chance to think about it: "When the bad outcome connected with a given response option comes into mind, however fleetingly, we experience an unpleasant gut feeling" (173). Similarly, when a *good* outcome is connected with a certain response option, this will also be sedimented in the body-brain and manifest within an internal preference system:

> somatic markers are a special instance of feelings generated from secondary emotions. Those emotions and feelings have been connected, by learning to, to predicated future outcomes of certain scenarios. When a negative somatic marker is juxtaposed to a particular future outcome, the combination functions as an alarm bell. When a positive somatic marker is juxtaposed instead, it becomes a beacon of incentive. (174)

Based on somatic markers, the body-brain makes decisions without our being conscious of it. Damasio uses the term "body loop" to describe the complex set of biochemical interactions or the somatically marked pre-sets of the affective wave, the specific neural circuitry that constitutes a certain "marker."

The fact that the machinery of primary emotions generates biochemical and mechanic bodily responses to a stimulus does not affect the assertion that "most somatic markers we use for rational decision-making probably were created in our brains during the process of education and socialization, by connecting specific classes of stimuli with specific classes of somatic state" (177). In other words, they are based on secondary emotions:

> Early in development, punishment and reward are delivered not only by the entities themselves [which later call for a response] but by parents and other elders and peers, who usually embody the social conventions and ethics of the culture to which the organism belongs. (179)

Somatic markers are not trained in one-to-one real-life situations but in the observation of interactions with those identifiable individuals who model behavior and regulate the child's actions through sanctions or rewards. Without having theoretical backing here, I think it is safe to say that at a certain phase of development, there is nothing more imperative than to "be a good one of one's kind" (Davis qtd. in Wynter, "Unsettling" 271). The child seeks to achieve this recognition by way of imitation (words, actions, mimics, gestures) even if it does not understand them in their *grown-up meanings*. Words, gestures, and so on thus attain an affective meaning without the linguistic differentiation an adult may (or may not) draw on. I believe this is what Claxton refers to in his example of a child attaching the somatic marker of disgust: "If Mummy makes a face of disgust when I start to investigate the food in the dog's bowl, I learn to attach my own disgust to similar sights and smells" (122). Even though this is a life-long process, the "critical, formative set of stimuli to somatic parings is, no doubt, acquired in childhood and adolescence" (Damasio 179), when we develop the words our mouth prefers to utter (Lyotard 2). This is when we *learn* our *culturally specific* subjectivity, enact it, find it, and maintain it. This is when the amygdala and the limbic system are formed most intensely. In this phase we learn deeply and our "internal preference system and sets of external circumstances extend the repertory of stimuli that will become automatically marked" (Damasio 179). It is also when our body begins *to language* by interpreting, applying, and making sense – hence the central role of children's books in Fanon's theorization. Every reaction we observe in our parents or peers towards any object, mode, or phenomenon racialized as Black will structure and predetermine our emotional reactions to Blackness as a white subject. Here I mean Blackness in all its forms: individuals, sounds, spaces, modes, and aesthetics. Just as we learn from "Mummy's face of disgust" (Claxton 122) not to touch the dog's bowl, we white people learn from daddy's joyful or sorrowful grimace when he practices the blues on his guitar most authentically, from our aunt's anxiously tightening her grip when a group of young Black men walk past, from the favorite cousin's relaxed face when he smokes weed and listens to Bob Marley or their anger when putting on Public Enemy. By observation, we develop an internal preference system that *serves us* to perform body states and behaviors we like, most explicitly in the praxis of anti-Black abjection as it leads to the "functioning balanced biological states" (Damasio 179), providing us white subjects with a subject-aeffect. Again, this highlights a culturally specific, distinct emotional pre-set, a more or less defined affective configuration of the body-brain's various dimension of anti-Black abjection as the visceral performance towards the anticipated somatic state of the subject-aeffect, which our consciousness may register merely as unraced homeostasis.[15]

---

15 Damasio writes that "the buildup of adaptive somatic markers requires that both brain and culture be normal" (177). This means that a somatic marker that aids social adaptability will necessarily be anti-Black. The somatic marker hypothesis thus helps to conceptualize how white culture ingrains whiteness in its subjects, how it naturalizes white abjectorship by inscribing it on all levels and regions of the body. Even though arbitrary and violent in its myriad manifestations, anti-Blackness is *normal*, ensuring the white supremacist culture of the modern West

Considering the subject-aeffect instigated by abjection as an emotion in Damasio's terms, as a psychobiological body loop, that can (potentially) be defined in terms of biochemical reactions and set apart from other body loops – whatever the precise emotional profile may look like – we can say that the moment of subject-making will register in and for the body-brain as a very distinct emotion – or a clearly definable combination thereof. Damasio identifies the five fundamental emotions as happiness, sadness, anger, fear, and disgust, each with variations like "euphoria and ecstasy" for happiness (149). We can assume that combinations of happiness and fear must also be possible – otherwise, "horror" as a combination of happiness and fear would not succeed. Without being in the position to detail the chemical, biophysical, or neurological specificities of a relatively complex body loop, we can still think about the specifically anti-Black, abjective subject-aeffect as an emotion in Damasio's sense. We might even go as far as to speculate that it has two (or more) levels. This way we can make sense of the fact that what we may experience as separate feelings (fear and desire) are superficially disparate, but draw on the same underlying sensation of experiencing a sense of self. Fear and desire, then, are two ends of the subject-making continuum in the white libidinal economy.

While cultural studies has long adopted psychoanalytic approaches, to my knowledge few neuroscientists have been interested in or conceptualized such a link. This appears to me grave neglect as it would help us to further materialize the kind of *culturally specific abjective subject-making phobia* (negrophobia) and the *culturally specific abjective subject-making desire* (negrophilia) I have elaborated here. Turning to Kristeva opens up this line of thought as well:

> Let me say that want and aggressivity are chronologically separable but logically coextensive. Aggressivity [...] merely takes revenge on initial frustrations. But what can be *known* about their connection is that want and aggressivity are adapted to one another. [...] Fear and the aggressivity intended to protect me from some not yet localizable course are projected and come back to me from the outside: "I am threatened." The fantasy of incorporation by means of which I attempt to escape fear (I incorporate a portion of my mother's body, her breast, and thus hold on to her) threatens me nonetheless, for a symbolic, paternal prohibition already dwells in me on account of my learning to speak at the same time. In the face of this second threat, a completely symbolic one, I attempt another procedure: I am not the one that devours, I am being devoured by him. (39)

It is worth pointing out that the "fantasy of incorporation" as a way of dealing with abject, phobic Blackness resonates with bell hooks's famous essay "Eating

---

remains *healthy*. (Despite the fact that we might describe that very culture as sick or pathological.)

the Other." Whether the individual in question needs a libidinal boost to conceive of themselves as fully Human (personally or socially), Blackness provides the portal to that very white jouissance. Even though the causal link is arbitrary and logically *wrong*, as white people, we are always bound to feel phobia/want in our encounters with Blackness qua abjection. Reading Blackness as a somatic marker means that Blackness (or the sphere of Blackness and whatever is abjected into it) refers to a floating signifier, a multidimensional yet elusive sign with no actual referent. Blackness itself does not denote anything and has no core, cultural or otherwise. Blackness gets attached to semiotic materialities that are not necessarily connected to each other on ontological grounds. Blackness is constructed by white people and attached to actions, motions, styles of music, activities, sounds, geographies, levels of income, gestures, fields of profession, types of sport, and ways of speaking, all racialized as Black. A white subject-body is the only locus where Blackness attains this abjective *meaning* in the form of the biophysical image or map that presents to consciousness the goings-on of the body in its encounter with Blackness. A brief and final look into Damasio's explication of the term somatic marker helps to illustrate this: "Because the feeling is about the body, I gave the phenomenon the technical term *somatic* state [...] and because it 'marks' an image, I called it a *marker*" (173).[16] Although we are looking at the effect of a signal on the body the affect attached feeds back into the real world. Our biochemical reaction to the sign-bearer *marks* that sign-bearer. Our *emotion* marks the external world. Blackness exists as a praxis of abjection because white libidinal economy relies on the never-ending abjective marking of what it construes as Black.

I suggest the oxymoron "solipsist encounters" to describe and denote the fact that a white subject cannot encounter a Black person without being thrown into some sort of solipsistic communication with themselves. White subjects' reactions have nothing to do with the Black-racialized individual standing in front of them and everything to do with the abjection necessary for their own subjectivity. These abjective perceptions, voluntary or involuntary, make any encounter with what is construed as Black a solipsistic event, inspired or triggered by an external stimulus that is actively and *abjectively marked as Black*. Given that

---

16 Note that the receiving body marks the external stimulus, which makes it an abjective *act*. This often gets unjustifiably twisted. It is essential to understand that a feeling a white individual has, involuntary as it may be, affects the external object. This relates to Freud's statement in "Claims" that "the principle function of the mental mechanism is to relieve the individual from the tensions created in him by his needs. One part of this task can be achieved by extracting satisfaction from the external world; and for this purpose, it is essential to have control over the real world" (186).

anti-Blackness is the ground on which we symbolically conceive of ourselves as subject-bodies and ushers us into the political sphere of the modern West, anti-Black abjection becomes the defining procedure of somatic marking if there ever was one. The theoretical consequences as they pertain to interpersonal encounters are unnerving. As white-bodied subjects, we cannot control all our affective reactions to what we read and mark as Black, including people who are structurally racialized as Black – whether they are close friends, passing strangers, or celebrities. The ultimate consequence of this theorizing is that any encounter with those who bear the Burden of Blackness (Fanon) involves an intuitive, abjective impulse that takes place within white subjects. And it will also, in one way or another, emanate from our white subject-bodies as well. These are the familial ties of whiteness. The continuum of anti-Black abjectorship encompasses Roy Bryant murdering and mutilating Emmet Till, white scholars engaging Black Studies theory, white gamers selecting Black avatars for their gangsta roleplay, singers in a white gospel choir, and innumerable other abjective agents (and myriad functional personas for individual white subjects) who constitute the collective subject-body of the modern West.

As discussed above, I read anti-Black abjection as an act of experience, as an inward psychic procedure with outward effects, and as an action that is internally motivated and feeds back into the psyche. This approach is not designed to relativize the physical violence directed against actual bodies compared to the white youth listening to hip hop. But it is designed to allow me to talk about the way that anti-Black abjection lies at the core of white subjectivity, and not only for those who explicitly fashion themselves as white supremacists. Damasio's by now classic and widely-accepted notion of the body loop as well as its vacuous version, the as-if body-loop, speak to that. Both loops mark an external stimulus somatically. Both serve to activate a previously developed biochemical "pre-set," to use Claxton's term. The difference is that the primal body loop denotes the idea of an actual, genetic predisposition of the human body – for example, being wary of falling from a height. In a situation where we are approaching a great height, our body might know and react to it before it has been relayed to our somatosensory cortices. This involves an actual change in the body before our *learned* brain can send the relevant information back into the body. In the as-if body loop, on the other hand, "the body is bypassed and re-activation signals are conveyed to the somatosensory structures which then adopt the appropriate pattern" (Bechara et al. 297; see also Damasio et al.). This means that in a given situation, the body does not give the information about the danger to the brain, but rather the *learned body-brain* knows what to do first. This is how, when we speak of somatic marking, we are in the sphere of what Damasio calls secondary emotions.

There is some danger in this transdisciplinary conflation of terms. The primal quality for Damasio addresses *actually instinctive* reactions to external stimuli. Assuming that anti-Blackness is genetically built into whiteness on this level would involve the actual, epistemic, and ontological naturalization of the Black-racialized body as inhuman on the grounds of culturally specific reactions to it. The psychoanalytic toolkit can, however, only be applied *within language*, even if the unconscious is assumed to exceed it. But this in no way disqualifies the connection; the results of both "the 'body loop' or the 'as-if body loop' may become overt (conscious) or remain covert (non-conscious)" (Bechara et al. 297). When talking about culturally specific abjection, then, we are within the sphere of the as if-body loop, which, as abjectors, we would not be able to differentiate from an actual body loop – both feel and are exactly the same when we notice them (overt somatic marking), and even when we don't (covert somatic marking). This means that, when reading anti-Black abjection as a secondary emotion, we must understand it in hybrid terms as a somatic marking of something real or imaginary, but observable as a physical act or experience, such as the perception or utterance of words. In *Words Can Change Your Brain* (2012), Newberg and Waldman argue for the power of language to materially, biophysically alter our brain structure, which corresponds directly to Sylvia Wynter's conception of being human as bio-mythoi hybrid practice. They argue that a word like "no" causes "a substantial increase of activity in your amygdala and the release of dozens of stress-producing hormones and neurotransmitters," which immediately interrupts "the normal functioning of the brain" (24). The youngest of the languaging Human bodies are thus prone to a learning experience that will sediment deeply in their bodies as affective knowledge: the more "negative thoughts they have, the more likely they are to experience emotional turmoil" (Newberg and Waldman 25). The more often they hear the word "no," the more vulnerable they are to clinical depression for the rest of their life. Newberg and Waldman also observe that "fearful words [such as] "'poverty,' 'sickness,' 'loneliness,' and 'death' [...] stimulate many centers of the brain, but they have a different effect from negative words." Rather than leading to depression, "the fight-or-flight reaction triggered by the amygdala causes us to begin fantasize about negative outcomes, and the brain then begins to rehearse possible counterstrategies that may or may not occur in the future" (25–26). These "fearful words" are somatically marked.

If a simple "no" can have this effect, how can we frame the power of the signifier "Blackness"? First, mobilizing Damasio's as-if body-loop and anti-Black abjection as inward (also imaginary) and outward (concrete action), we must remember that the brain "does not distinguish between fantasies and facts when it perceives a negative [or other] event" (Newberg and Waldman 24–25).[17] Second, our active performance strengthens whatever affective sensation a word provides for us: "If you vocalize your negativity, even more stress chemicals will be released, not only in your brain, but in the listener's brain as well" (Newberg and Waldman 24).[18] Third, we must remember the evolution or development of our semiotic capabilities: "Before we learn to think in words, we instinctively think in pictures. As the brain continues to develop, we gain the ability to think in increasingly abstract ways," namely from picture to drawing to symbol to word, as Newberg and Waldman suggest (49). All of this is highly consequential when related to anti-Black abjection. Imagination, memory, actual encounters, fictional representation, factual representation – the list goes on – are all able to provide anti-Black affective jouissance. Adjudicating whether one mode is more powerful or efficient than another would be mere subjective speculation. Further, we must follow Broeck's assertion that abjection is primarily an act. If we utter anti-Black language (or perform outward anti-Black abjection in another way onto the world), it enhances our white sense of self, providing us with a white subject-aeffect. Finally, the whole procedure does not even need a physical manifestation of Blackness. There are "some words our mouth prefers to utter" (Lyotard 2) even if we white people do not understand why. While Newberg and Waldman are primarily interested in word language, I think we must also read any semiotic unit involving gestural, behavioral, sonic, or other in the meanings of Blackness – and all their real-world manifestations.

Hybrid mode: the non-location of abjection in bios-mythoi

I want to pause here to consider the relevance of Kristeva's concept of abjection in relation to racism. What do we make of the fact that her notion of abjection is *in itself* an elaborate manifestation of the modern matrix, given that it too assumes that we *naturally* feel disgust for physicality, fleshliness, the corpse? Can and should we consider it a workable concept for a critique of anti-Black-

---

[17] There are even studies showing that the brain does not distinguish between physical and mental pain (see, for example, Eisenberger or Kross).
[18] At this point, Newberg and Waldman consider words associated with negativity. However, in other sections, and in keeping with the general thrust of their book (which is mostly a self-help guide for communication and self-improvement), they quote from studies that demonstrate the positive effects that "positive" words have on the brain, even developing training techniques out of this fact.

ness when it can so easily be mobilized to *naturalize* in ontogenetic terms? If so, where exactly do we locate the abjective process in Wynter's conception of being human as bios-mythoi hybridity? Martina Tißberger's criticism of Kristeva's abjection as mobilized by Derek Hook is constructive:

> Hook wants to explain racism with recourse to Kristeva's ontogenetic construct, which is in itself already *prefigured* by that very racism. It all fits together so nicely because one is the model for the other. Hook fails to take a closer look at abjection as an ontogenetic moment. He describes the infant's abjective process as "existential," "ontological," "natural," "essential," and "instinctive" in the struggle for "individual coherence," "wholeness," "identity." We learn nothing about why coherence and wholeness are so important that they are "produced instinctively" and have ontological quality. It seems natural from the Western perspective in a culture that values individuality and autonomy while judging collectivity and interdependence negatively. However, beyond the Western, hegemonic point of view, "coherence," "wholeness," and "identity" do not apply to the foundations of culture and subject. (48–49, my translation)[19]

Hook's discussion and mobilization of abjection as it relates to racism, and his concept of "'pre-discursive' racism," are not my reference for reasons in part considered in Tißberger (compare Hook, "Racism as Abjection" and "Prediscursive Racism"). However, I am aware that much of her criticism may also be directed against this project, and has remained unaddressed to this point. In itself and perhaps regarding its immediate object of critique, her argument has great validity. However, Tißberger's criticism is based on several axioms I do not share. "Taking a closer look at abjection as an ontogenetic moment," for example, is only imperative for those who aspire to an ontogenetic truth about humanity at all, focusing on the actuality of abjection for those who experience it. This is the crux of Tißberger's otherwise significant argument. The moment we think about and apply abjection as a *tool to talk about an observable reality* rather than as a vehicle that takes us to an (ontogenetic) truth, we are doing something else.

---

19 "Hook will im Rekurs auf Kristeva's ontogenetisches Konstrukt den Rassismus erklären, der bereits Kristeva's Konstrukt präfiguriert. Das alles passt so schön zusammen, weil eines des anderen Vorbild ist. Hook versäumt, Abjektion als ontogenetisches Moment genauer zu betrachten. Er beschreibt den Prozess der Abjektion beim Kleinkind >existenziell<, >ontologisch<, >natürlich<, >essenziell< und >instinktiv< im Kampf um >individuelle Kohärenz<, >Ganzheit<, >Identität<. Wir erfahren nichts darüber, warum Kohärenz und Ganzheit so wichtig sind, dass sie >instinktiv< hervorgebracht werden und ontologische Qualität haben. Es scheint selbstverständlich aus der westlichen Perspektive und ihrer Kultur, die Individualität und Autonomie wertschätzt, während sie Kollektivität und Interdependenz negativ bewertet. Jenseits des westlichen, hegemonialen Standpunktes gelten >Kohärenz<, >Ganzheit> und >Identität< jedoch keineswegs als Grundfesten von Kultur und Subjekt" (48–49).

3.3 Embodying anti-Black abjection — 109

In consequence, the argumentational circularity Tißberger points to – suggesting that abjection explains racism so aptly because it is modeled on that same racism – is not so flawed as it might appear. Again, taking abjection as a tool to talk about a reality that can be described as structurally abjective opens up a register otherwise foreclosed to the analyst *beyond* the culture-nature dichotomy. Considering abjection as a primal, biological experience of entering discourse *as well as* one created and perpetuated by discourse may be circular, but it is not therefore false. If we think of it in Wynter's conception of being human, "hybridity" becomes key:

> Once you redefine being human in hybrid mythoi and bios terms, and therefore in terms that draw attention to the relativity and original multiplicity of our genres of being human, all of a sudden, what you begin to recognize is the central role that our discursive formations, aesthetic fields, and systems of knowledge must play in the performative enactment of all such genres of being hybridly human. You will begin to understand, in the case of the latter, that the role of all such knowledge-making practices with respect to each genre is not to elaborate truth-in-general. Instead, the role of such knowledge-making practices is to elaborate the genre-specific (and/or culture-specific) orders of truth through which we know reality, from the perspective of the no less genre-specific who that we already are. These genre-specific orders of truth then serve to motivate, semantically-neurochemically, in positive / negative symbolic life / symbolic death terms, the ensemble of individual and collective behaviors needed to dynamically enact and stably replicate each such fictively made eusocial human order as an autopoietic, autonomously functioning, languaging, living system. ("Catastrophe" 32)

To me, this speaks to the use of abjection on two levels. Looking at abjection in its ontogenetic dimension, which Tißberger suggests is worthy of further consideration, I take from Wynter that we can read abjection as hybrid, both nature and culture at the same time. Not as one or the other, not as a switching or oscillating, but actually as one and the same – a hybrid existence that precisely does *not* differentiate between either. If we conceive of being human in bios/mythoi hybridity, there is no need to qualify whether we should view abjection as a linguistic, racial concept within a sphere of discourse that exists before the (allegedly non-raced) infant can abject, or as an (allegedly) naturalized part of a general humanity's condition articulated through the modern matrix that also prefigures the grounds of anti-Blackness (because it has been itself developed on those grounds). The element of hybridity in Wynter's conception makes it difficult to fathom in its entirety because it challenges the chronology of causality and logic that Tißberger applies to denounce the explanatory capacity of abjection as a critique of anti-Blackness. Where Tißberger criticizes the racial prefiguration of the concept as developed by Kristeva, following the urge to *locate it in its ontogenetic existence for a general humanity* at all is a fantasy of un-raced transcendence on the part of the scholar. Rather than being *located somewhere*, (anti-Black) abjection should be put to use as conceptual practice in terms of sociogeny, of eusociality.

Tißberger's most poignant criticism is that we can only talk about a subjectivity generated via "individual coherence," "wholeness," and "identity" when talking about abjection. This may be true, but the fact that these notions pertain to culturally specific Western concepts of subjectivity does not demand we renounce their power to explicate that very subjectivity. Any explication will certainly fall short for an elaboration of a generally racial tendency of a universally-conceived humanity – whatever that would look like and how (or why) one would wish to arrive at it – but it can be mobilized to consider the culturally specific anti-Blackness of the modern subject: the improvising subject as Man1 who "overrepresents itself as if it were the human itself," to use Wynter's phrase ("Unsettling" 260). We may not learn "why coherence and wholeness are so important that they are 'produced instinctively'" (Tißberger 48), but we can take these as facts regardless of their ontological quality. Also, with Broeck, we can use its capacity to give language to actions in terms of language and affect done by white subjects, who are abjectors by definition. While it may appear circular to critique the white modern subject *from within*, it is still the only available language this subject may understand. Wynter posits with Césaire "the study of the Word/mythoi" ("Catastrophe" 32) as a way to understand not only ourselves but our bios. Mythoi become an external ground from which we can see how our biological being exists. The language we use to discuss ourselves provides a way to understand ourselves. Through the study of the Word, we can avoid projective academic tendencies into nothingness (assumptions about generic and supracultural Human ontogenetic existence).

This conceptual bracket must be borne in mind through the rest of the project. When working with Winnicott's theory of play, for example, I reframe his conception of transitional objects as transitional Blackness, providing a culturally specific white jouissance of abjection and creating a subject-aeffect. This enables me to make statements about white subjects playing, and about why they are predominantly white in the case of improv. (To talk about Black absence in improv by looking at assumedly "ontogenetic" characteristics of Blackness is far from my mind.) Similarly, with Freud's theory of humor, I discuss the anti-Black violence of white laughter rather than laughter in general, and rather than analyzing Black cultural productions of humor. I have no interest in dissecting or analyzing those who do not hold white modern subjectivity, or speculating about whether abjection exists for them. I make no statements or assumptions about what might trigger a subjectoid-aeffect in those bodies precluded from what modernity knows as subjectivity or what their brain scans might look like. I do not apply abjection in a generally humanist project. I do not mobilize it to talk about people racialized as Black.

3.4 Modern popular culture as Blackness

Many scholars have considered, analyzed, and thereby performatively constructed a *relationship* between cultural production imagined as distinctly Black and an otherwise unmarked popular culture. The concept of cultural appropriation looms large in this academic field, and we clearly need a strategic and broadly applicable term to talk about how white people capitalize on (their performances of) Blackness. In the white supremacist system, white people emulate Black performance to become the Kings of Swing, Jazz, or Rock and Roll, capitalizing on styles, modes, sounds, and forms developed by artists racialized as Black. At first glance, this calls for the analytical tools of appropriation and authenticity. I do think that relevant, strategic, documentary, archival, and practical work can be done with these terms. A lot of academic labor has gone into the critique of white people capitalizing on Black forms and how Blackness has been *misrepresented* in popular culture. However, the rhetoric of appropriation cannot capture the foundational (libidinal, historical, structural) functions of (anti-)Blackness in what we know as a generalized popular culture. I argue it is ill-fated to consider popular culture writ large as a space for negotiation in which the appropriation or misrepresentation of Blackness can even take place. No "relationship" exists between Blackness and popular culture because modern popular culture is *sui genesis* the result of white abjectorship and the violent maintenance of a *non*-relation. The rhetorics of appropriation, misrepresentation, and authenticity interfere with a clear view of this structural ground. Virtually every aspect of global Western popular culture can be traced back to the sphere of Blackness as it comes about qua anti-Black abjection. Below, I consider the default lines of argument about appropriation and misrepresentation, showing that they fail to address the foundational anti-Blackness of popular culture. Then I turn to Saidiya Hartman to argue for a historicized view of the anti-Blackness of popular culture before I analyze it as a trading space for white affective stimulation through culturally specific modes of modern anti-Black abjection, engaging with Adorno's infamous jazz critique as an example.

Notes on the rhetorics of appropriation

Appropriation is defined as "the process whereby members of relatively privileged groups 'raid' the culture of marginalized groups, abstracting cultural practices or artifacts from their historically specific contexts" (Dines and Humez 567) or the "taking – from a culture that is not one's own – of intellectual property, cultural expressions or artifacts, history or ways of knowledge," as in the 1992 resolution of the Writer's Union of Canada (qtd. Ziff and Rao 1). The language of appropriation always involves some sort of theft, an unlawful and unequivocal taking. Accordingly, the whole discourse around the term is ultimately one of property and property rights as they play out between cultures, about who has those rights and how their violation can be articulated or even sanctioned. Such an approach presupposes a Herderian understanding of culture, which relies on concepts of plurality and relations between cultures. Such relational plurality, however, cannot describe the non-position of Blackness within the white modern matrix. Despite their strategic workability, arguments based on appropriation fall short in articulating the complex and more fundamental issues at stake. Emerging from a romantic fiction of cultural relationality, they cannot provide sufficient traction to address the anti-Black popular culture at large. Debates about appropriation are at best superficial, at worst complicit in covering up the fundamental anti-Black abjection that constitutes both the historical ground and contemporary practice of popular culture.

To discuss this in more detail, a closer look at the logic of the appropriation discourse is helpful. Susan Scafidi theorizes cultural appropriation within the legal terms of intellectual property law because both presuppose an idea of intangible property. She assesses a "legal vacuum" when it comes to cultural group creation, whereby some forms of "creative [intangible] production receive extensive, *even excessive*, protection against copying under our system of intellectual property law," namely those that safeguard individual (intellectual) property. At the same time, cultural products are generally "indefinite works of group authorship, and they represent a particular challenge" for lawmakers (11). The allegedly paradoxical "legal vacuum" between individual and collective property rights is easy to understand in view of what constitutes a "group" within the matrix of the universalist global West, because it begs the question for whom respective property rights count. While laws are in place to protect the private and intangible ownership of individuals and privately-owned corporations, the same cannot be said for communal production or the collective ownership of intangibles such as cultural materials, knowledges, or styles. I suggest, therefore, that what presents itself as a "legal vacuum" in Scafidi's legal studies approach speaks to the *lack of property relations* between Western subjects and non-subjects who inhabit the sphere of Blackness – and not the lack of a legal grammar to address it. Departing from Scafidi, I believe that the legal and executive demarcation lines are not drawn between the individual and the collective. Rather, they are drawn between subjects with the capacity to own and those racialized as Black, who are thereby owned and constitute fungible, negotiable material to be accumulated (following Wilderson).

Echoing Wynter's critique of how modern Man "overrepresents itself as if it were the human itself" ("Unsettling" 260), the authors of "A Broken Record" remind us that Western culture is invisible to itself, despite being one culture among others:

> The point about [Western] Culture is it is cultureless; its values are not those of any particular form of life, simply human life as such. The universalist self-understanding of Culture puts it in a relationship of superiority to 'mere cultures' as blatantly historical forms of life that value collective particularity. (Coleman et al. 180)

Misrepresentation and stereotypification are common critiques against practices understood as cultural appropriation. In the preface to *Soul Thieves,* Tamara Brown writes that "misrepresentation refers to deliberate, typically negative, depiction of a false ideal" commodified for capitalist gains (viii). The reproduction of negative stereotypes for capitalist gains comes across as ethically untenable. But as Scafidi's concept of an "identity tax" suggests (7), it also brings along the welcome idea of a cultural core, a communal source, a culture that can be violated in the first place. We can *distort* Italians by stereotyping them as Mafia bosses, poke fun at Germans by portraying them wearing white socks and sandals, and we can even produce derogatory images of First Nations because we think they have a particularly "savage" yet "noble" core. Such violations rely on a cultural locus from which claims can be made in defense of cultural material or even cultural truth. These gestures assume a potential authenticity, implying that there is a right way to represent a culture. While the law may not provide the instruments to sanction this sort of appropriative misrepresentation, an ethical code still applies. Here the notion of authority qua authenticity comes into play, which "may thus compensate for an inability to secure or protect ownership of an embodied idea, creation, or design" (Scafidi 53). Scafidi suggests that "in the unregulated, intangible world of cultural products, unenforceable assertions of ownership can instead *take the form of* 'authenticity'" (53), or that "the *rhetoric* of authenticity performs much the same social function as property ownership, placing the claimant group in a position superior to all others with respect to the item in question" (54). Authenticity claims are thus *like* property claims without their political and financial ramifications. Scafidi finds the idea of authenticity "useful" because it "establish[es] source communities as the definitive repository of cultural meaning with respect to those [cultural, intangible] products" (53).

The notion of an authentically representable "source community" is the first instance where the rhetoric of cultural appropriation does not hold up in view of Blackness. Blackness does not and can never denote an authentic cultural source, always being open and vulnerable toward the white abjection that brought it about in the first place. The sliding signifier of Blackness does not denote a cultural core that the source communities envisioned by Scafidi can reckon with, which makes the "authentic representation" of Blackness impossible. Even recognizing it as a misrepresentable, appropriable "mere culture" is foreclosed to those racialized as Black. For abjected Blackness, there is no cultural core, no origin, no telos that can be envisioned *within the matrix of white supremacy*. Theorizing through the lens of intellectual property, Scafidi maintains that through its "Enlightenment parentage, it inherits a tremendous confidence in the ability of the rational mind to create, to solve, to progress, to assign value" (11). The absence of a cultural core inhabited by subjects makes it logically impossible to produce intellectual property from within the sphere of Blackness because neither intellectual capacity nor cultural authenticity is believed to reside there. The authors of "Broken Record" recognize the discursive function of "European intellectual property doctrines":

> In the rhetoric that legitimates intellectual property, the aesthetic work, be it literary, artistic or musical, both embodies the personality of its individual creator and makes a singular contribution to human civilization, universally conceived. (Coleman et al. 180)

The idea of intellectual property is thus in itself raced. It restricts the realm of the appropriable to those with the capacity to produce and therefore to lose something in the first instance. Scafidi deals with "the intangible aspects of creations of the human mind" (14). Accordingly, the realm of the appropriable is restricted to those with the discursive capacity of performing a *Human mind*, that is, those groups that comprise what Wilderson calls the "race of Humanism," which is defined *ex negativo* by everything that is not Black, and by everything that the Black is not (20).

In terms of intellectual property, this means that there is nothing to appropriate because there is no locus of ownership. There is no space to be *invaded*, no culture to be *raided*, no individual's property rights to be *violated* – which does not mean there is no violence involved, but that there is no individual or collective subject against whom such violence qualifies as transgression. Consider this long excerpt from Nicholas Brady's article "Looking for Azealia's Harlem Shake:

Or How We Mistake the Politics of Obliteration for Appropriation" on his blog *out of nowhere:*

> Appropriation depends on defining our relationship to objects through the lens of property relations, so that an object is the property of a person or group. This relation is always already thorny but is especially cut by cultural objects. Cultural objects can certainly be commodified, but the issue of ownership is always wrapped up in relations of power, privilege, and propriety [...]
>
> If we are to talk about commodities and property in relation to culture, this should swerve us face-first into the topic of slavery and specifically the "human commodity" known as the slave [...]
>
> The most horrifying to consider here is that the very happiness of the slave was owned by the master – this means the master often forced the slave to perform songs on the auction block, in the coffle to it, and for the slave to smile and laugh and joke in his/her presence (this is described as the "terror of pleasure" by Saidiya Hartman in her magnum opus *Scenes of Subjection*). This politics of appropriation can find its "origin" dispersed among the performance of domination we know as the peculiar institution. What we have called "appropriation" implies that black people own their culture and the master stole it from them. Yet, when we let go of romantic terms our claim sounds like this: a piece of property owns a piece of property and was stolen by the citizen who owns them both. How does a commodity own a commodity? How does the owner of that commodity steal a commodity from his own property? [...]
>
> All that is to say that the concept of appropriation mystifies what is actually happening when white people "steal" black culture. Stealing implies a crime or a sense of wrongdoing or doing something improper. Yet the very concept of the proper – as well as property – depends on the black to be radically open to violation. So it is not improper to violate the black, it is in fact the definition of the proper itself.

Brady succinctly defines the radical consequences of the ways in which proprietary lens "skew[s] our ethical considerations" (Coleman 175). He mentions Hartman's *Scenes of Subjection* to discuss the flawed concept of appropriation in popular culture. I take his cue and read her multidimensional magnum opus as a call to historicize popular culture as always already Black and anti-Black at the same time.

Historicizing the popular I: the auction block as the primal scene of US entertainment

"[W]e are left to ponder," Saidiya Hartman writes, "whether the origin of American theater is to be found in a no-longer-remembered primal scene of torture" (*Scenes* 32): the coffle, the handling center of the enslaved and the space in which Blackness was created for a Euro-American commons. By trading Black bodies, the modern racialized order was performed through the materialization of who can own and who is owned. In the coffle, the racialized dehumanization of Black bodies was actualized qua performance.[20] Yet the coffle was not only the space where Blackness was spectacularly theatricalized. It was also a theatrical event in its own right. Enticed purely by the entertaining qualities of the spectacle as such, an audience would show up with no interest in buying an enslaved body (or the financial resources to do so). The auction block was Euro-America's first popular theater stage. Even though it may function as one, this is not a metaphor. The aesthetic configuration of the coffle includes everything a theatrical situation demands: a stage, human-objects performing for an audience, costumes, director's instructions. Thanks to Hartman's immense archival work, there is no question about the entertaining amusements facilitated by and performed in the coffle. It is worth recognizing that the whole event was the performance of communal jouissance. The act of the cheerful Slave Jim Crow was only one element – the object of performance, so to speak – of the *generally* "festive atmosphere of the trade [that] attracted spectators not intending to purchase slaves" (*Scenes* 37). The auctioneers were no dry salesmen but entertainers themselves, "clown[s] [who] made funny talk and kept everybody laughing" (Carter qtd. in Hartman 37). The Black-racialized enslaved performers were given clothes by the stage-managing coffle driver "just before they reached the market space," who retrieved them after the show was over: costumes (Blassingame qtd. in Hartman 38). The theatrical presentation of the enslaved at times involved curtain-drawing and the presentation of musical skill or dancing abilities – both performed under the coercive direction of the whip. In its aesthetic and its political configurations, this was a casting in which white enslavers put together their ensemble to reenact white plantation fantasies for everyone, a communally shared event of staged anti-Black abjection. The organizers took advantage of the linkage between entertainment and selling refreshments, capitalizing on the anti-Black festivity as much as on the actual trading of Black bodies: "The distribution of rum or brandy and slaves dancing, laughing, and generally 'striking it up lively' entertained spectators and gave meaning to the phrase 'theater of the marketplace'" (Hartman, *Scenes* 37). It would thus be a mistake to think that the entertainment dimension was in any way secondary to the selling of the enslaved. While the business was the occasion, a good show was central to a vendor's success. Hartman quotes an 1853 article from the *New Orleans Daily Picayune*:

---

20 Here, I focus on the coffle's function as a reenactment of the original dehumanization that performatively actualizes that idea for the settlers. For the enslaved, this process started in their deportation.

> Amusements seldom prove attractive here unless music is brought to the aid of other inducements to spend money. So much is this the custom and so well is this understood, that even an auctioneer can scarcely ral[ly] a crowd without the aid of the man with the drum. (*Scenes* 38)

Euro-American encounters with enslaved Africans as presented on the auction block were theatrical, which gave birth to the idea of an all-American theater in "the theater of the marketplace that wed festivity and the exchange of captive bodies" (Hartman, *Scenes* 37). Here lie the origins of US American theater, popular or otherwise – and of the specters that haunt popular culture even today. The spectacularized staging of anti-Black abjection later transformed into the aesthetic theatricality of minstrelsy: shows performed by Black artists or white artists in blackface. Many scholars point out that minstrelsy founded US popular culture, especially in the fields of theater and comedy, and constructed Black figures for centuries to come:

> Even as minstrelsy waned in popularity during the late nineteenth century, its impact continued to be felt. Blackface comics gradually disappeared from the American stage, but the *stage* black remained indelibly etched into the American mind until well in the twentieth century. (Watkins 102)

Hartman's focus does not lie exclusively with the dehumanizing aggression of those who killed and sold Africans, but with their audience. She conceptualizes the simultaneity of terror *and* pleasure of the auction block spectacle. In the initial chapters of *Scenes of Subjection*, she addresses how white subjects' alleged *empathy* for Black suffering facilitates musings on a generalized Humanity understood in the logic of its descriptive statement, which predetermines the subsequent symbolic and physical dehumanization of the Black bodies that enabled those musings. She considers abolitionist writing by John Rankin, an observer shocked by the goings-on at the coffle, and Abraham Lincoln on encountering a coffle on a steamboat. In fancying himself and his family in the position of those enslaved, Rankin seeks to make Black suffering legible by way of "facilitating an identification between those free and those enslaved" (18) in order to "make their sufferings our own" (Rankin qtd. in Hartman 18). In the very moment the white American enslaving subject extends their Humanity to what they know as a Slave, they are imagining a solipsistic endeavor through the signifier of Blackness – in this case their own fear of experiencing treatment "like a Slave." Pointing out how "in making the other's suffering one's own, this suffering is occluded by the other's obliteration," Hartman asks us to "consider the precariousness of empathy and the thin line between witness and spectator" that characterizes Rankin's abolitionist agenda:[21]

> Rankin must supplant the black captive in order to give expression to black suffering, and, as a consequence, the dilemma – the denial of black sentience and the obscurity of suffering – is not attenuated but instantiated. (*Scenes* 19)

Another example is Lincoln's writing on his encounter with a coffle on a steamboat. Ruminating on the Human condition *as such*, Lincoln is confused by the apparent cheerfulness of the enslaved. He cannot fathom how "they were the most cheerful and apparently happy creatures on board" (qtd. in Hartman 34). Projecting his universalized idea of Humanity onto them, those in shackles seem to live in a paradox: "either their feelings seem unwarranted considering their condition [...] or this proverbial cheer especially suited them for enslavement." For Lincoln, "the elasticity of blackness [that] enables its deployment as a vehicle for exploring the human condition," provides the jouissance of reasserting his own full white Humanity at the peak of civilization (Hartman, *Scenes* 35). Empathy functions here as the white modern subject imagining themselves as a morally or ethically "good man or woman of one's kind" (Davis qtd. in Wynter, "Unsettling" 271), somatically ignorant of the fact that this specific kind has been brought about sociogenically by the same abjective praxis, and is grounded in the very matrical DNA that grants the white subject the capacity to perform such musings in the first place. Extending empathy in this way is to partake in the "overrepresentation of Man as if it were Human" (Wynter, "Unsettling" 267). However well-intentioned, such acts of abjective empathy are ultimately violent. Hartman mentions the term "abjection" only in passing, yet her argument resonates with Kristeva in that she reads terror *and* pleasure as drawing on the anti-Black matrix that creates a dehumanizing sphere of Blackness. These modes of abjection perform the same anti-Black sentiment and obliteration, ultimately uniting abolitionists and apologists of slavery.

---

21 White improviser Schleelein observes it is "such a fine line" for a white improviser to find a context in which the use of the n-word might be acceptable. Following Hartman, we can state that toeing the fine line that differentiates the abjector-as-witness from the abjector-as-spectator (both in action and perception) demarcates the start and end of white morality. In consequence, as a moral limit, this line also maintains morality in the first instance. In view of what I have been delineating as anti-Black abjection, this line is a discursive construct, which in performative practice is so thin that it nudges infinitely toward zero, towards nonexistence. Debating this line, asking what is right or wrong for a white subject to say or do, or performing regret that whatever a white subject does is always *bad*, maintains the moral stability of white supremacy.

Analogically, white abjectorship gives rise to two different genres of *Americanist* theater aesthetics, both equally anti-Black: melodrama and minstrelsy. Hartman writes:

> Despite differences between their respective conventions and stylistic devices, the uses made of the black body established continuities between minstrelsy and melodrama that surpassed their generic differences. Although the ethical valence of such violence differed, it nonetheless delivered a significant pleasure. Blows caused the virtuous black body of melodrama to be esteemed and humiliated the grotesque black body of minstrelsy. Uncle Tom's tribulations were tempered by the slaps and punches delivered to Topsy. The body's placement as ravaged object or as the recipient of farcical blows nonetheless established a corporeal language that marked Zoe, Tom, and Topsy as identifiably black and exposed the affiliations between the auction block and popular theater [...] Melodrama presented blackness as a vehicle of protest and dissent, and minstrelsy made it the embodiment of unmentionable and transgressive pleasures. (*Scenes* 26–27)

We need to understand (anti-)Black performativity within this historical trajectory, recognizing that anti-Black abjectorship transcends genre. We also need to recognize that anti-Blackness constitutes the origin of US popular culture as such – the culture that was to become globally dominant in the centuries to follow. There is no popular culture disconnected from this history, and there is no *subversive* position in popular culture that one may voluntarily chose to be "on the right side" of:

> The terror of pleasure – the violence that undergirded the comic moment in minstrelsy – and the pleasure of terror – the force of evil that propelled the plot of melodrama and fascinated the spectator – filiated the coffle, the auction block, the popular stage, and the plantation recreations in a scandalous equality. (Hartman, *Scenes* 32)

Recognizing the roots of US popular sentimental, humorous, and representational entertainment right there on the auction block and in the audience is consequential for contemporary scholarship on popular culture. If we consider popular culture structurally, as I propose, we need to think of it in its generic affective functionality rather than its particularistic meanings. Rather than seeking to *decode* Blackness *within* the grammar of unmarked white popular culture, I suggest we read popular culture *as* anti-Blackness in its aim to mobilize a certain affective (abjective) reaction *for white subjects*, namely the visceral wave, the experience of specifically anti-Black homeostasis developed above with Broeck and Kristeva as the culturally specific modern white subject-aeffect. However critically or romantically one may approach popular culture, it remains inextricably linked to its origins on the auction block, historically, libidinally, aesthetically, and financially:

> The relation between pleasure and possession of slave property, in both the figurative and the literal senses, can be explained in part by the fungibility of the slave – that is the joy made possible by virtue of the replaceability and interchangeability endemic to the commodity – and by the extensive capacities of property – that is, the augmentation of the master subject through his embodiment in external objects and persons. [...] Thus the desire to don, occupy or possess blackness or the black body as a sentimental resource and/or locus of excess enjoyment is both founded upon and enabled by the material relations of chattel slavery. (Hartman, *Scenes* 21)

Historicizing the popular II: lynching parties and the spectacular nature of Black suffering

+++TRIGGER WARNING+++ To further discuss the racial extremes to which the libidinal structure of anti-Blackness leads and how it functions through the popular, in this section I engage with Baldwin's short story "Going to Meet the Man." This literary piece includes several graphic descriptions and racist utterances by the white abjector in the narrative. Some descriptions are reprinted here to show how such (perceived) extremism relates to what is otherwise deemed harmless. I am aware that this intentionality does not circumvent the performative reiteration of reprinting; hence the trigger warning. +++TRIGGER WARNING+++

From reading the auction block as the primal scene of US popular entertainment and spectacle, it is a short step to the entertaining qualities of lynching parties for white audiences well into the 1940s. The documentation and theorization of lynching and lynching parties in white supremacist culture have been prolific. I am concerned with its performative dramaturgy and entertainment quality, which have only been considered more recently. Contemporary theorization no longer reads lynching as a historically distinct cultural phenomena of the past but understands that its fundamental structures are still very much alive. In *Legacies of Lynching*, John Markowitz reads the practice in its function to create what he describes with Halbwachs as "collective memory" (xxi) from which a "cultural repertoire" (xxii) derives. He notes that before the Civil War, the term "lynching" denoted "a variety of forms of punishment, including beating, whipping, tar-and-feathering, and, only occasionally, killing" that mobs of American patriots inflicted upon "Loyalists and British sympathizers" (xxiii). In the antebellum years, "mob violence became increasingly widespread and was directed against abolitionists, Mormons, Catholics, and blacks." The primary victims at the time were white people who "held unpopular moral or social beliefs or who engaged in behavior that was deemed inappropriate" (xxiii). Anti-Black lynchings were rare exceptions, for example in response to rebellions of enslaved

Blacks, or instigated by a *general* "extreme white insecurity" (xxiii). This rarity was not the result of recognizing the Humanity of Black-racialized people, but rather because lynching would have been conceived as property damage: "Slaveholders had financial or political stakes in protecting their slaves from lynchings." During and after the Civil War, however, "lynching" came to signify "putting to death" and was understood "'almost exclusively' as a method of punishment for newly enfranchised African Americans" (xxiii). Prior to the Civil War, then, lynchings functioned in the political realm, motivated by notions of law or morality. I suggest that when the practice became exclusively directed against formerly enslaved Blacks, such ethical or political negotiability loses ground, and the libidinal and symbolic functions take over to maintain the political sphere *as such*.

After considering traditional interpretations of what lynchings did and why they existed, Markowitz writes that "any explanation of lynching needs to account not only for explicitly political factors but also for economic and cultural motivations." The latter speaks to the notion of lynching parties as white, subject-making entertainment; parties varied in size but could grow into "massive public spectacles" that drew "thousands of participants" (xxv). Further, he lays out how lynching events must be read not only as spectacles in themselves but through their public announcements and documentation as well, through the selling of postcards (merchandise),[22] journalistic coverage, and distribution through rapidly-developing technologies to reach even more people:

> The specific form that spectacle lynchings took became routinized over time. The standard sequence of events included a hunt for the accused, the identification of the captured African American by the alleged white victim or members of the victim's family, the announcement of the upcoming lynching, selection of the site, and the lynching itself, which involved torture and mutilation, often including castration, followed by burning, hanging, shooting, or a combination of all three. (xxvii–xxviii)

There were also "special trains to bring spectators to lynching sites that had already been announced, and they would occasionally advertise these trains in local newspapers" (xxvi). Lynchings ran on the collective libidinal fuel that drove the economy, politics, and entertainment industry. Markowitz argues that lynchings on the whole "were intended to create collective memories of terror and white supremacy" (xxvi). While this hints vaguely at the anti-Black structures of a given culture and its libidinal ground, speaking of intention endows

---

22 Many of these postcards have been collected and published James Allen in *Without Sanctuary: Lynching Photography in America* (2000).

the organizers of these events with an agency and control over social meaning that I do not follow. Conceiving of these events as individually "intended" to keep Black-racialized people "in their place" in one way or another is pointless at best and dangerous at worst because it ignores the fact that it exists in the specific anti-Black configuration of an already existent libidinal structure. Reckoning with an active political or cultural intention requires the notion of a potential subjectivity for those racialized as Black. However, no such subjective position exists in the modern formations of sociality for Black bodies. Accordingly, it is less the scholarly classification of the functions and motivations of lynching parties than their literary treatment that takes us further here.

"Like a far-away light:" Baldwin's "Going to Meet the Man"

In "Going to Meet the Man," James Baldwin tells the story of a white deputy sheriff named Jesse, whom the reader first encounters lying next to his wife in bed at night. Jesse is sexually aroused but unable to have an erection, "silent, angry, and helpless," filled with excitement which "refused to enter his flesh" (229). Baldwin narrates a series of memories that lead Jesse further into his white supremacist anti-Black libido. It is a powerful piece, and I can only scratch the surface of all the meanings it entails, vertical, horizontal, and lateral. In this context, I read it as a psychological analysis, a discussion of the libidinal structure in which lynchings make meaning.

One of the remembered events is a lynching party Jesse attended with his parents when he was a young boy. Baldwin stresses at various points the festive and entertaining atmosphere of the event; the group they were going with looked "excited and shining" in anticipation, and they "were carrying food. It was like a Fourth of July picnic." As a celebration of national identity, the lynching event also provided an occasion for white people to meet and greet each other, not unlike the function of Sunday church. Jesse's mother "wanted to comb her hair a little and maybe put on a better dress, a dress she wore to church" (242). For Jesse, the whole family event and anticipatory sensations provided a feeling of comfort: he "got into the car, sitting close to his father, loving the smell of the car, and the trembling, and the bright day, and the sense of going on a great and unexpected journey" (243). His deeply ingrained and multidimensional comfort would from then on be linked to anti-Black abjection. The reader then follows Jesse through the event and learns about the harmonized affect generated in the crowd as they watch the burning of the Black body: "Those in front expressed their delight at what they saw, and this delight rolled backward, wave upon wave, across the clearing, more acrid than the smoke." (245) The crowd roared "as a man stepped forward and put more wood on the fire" (246).

When, before castrating the victim, "one of his father's friends" caressed and played with a knife "brighter than the fire [...] a wave of laughter swept the crowd" (247).

Given this account, we can no longer understand lynching parties only as the vigilante projects of white supremacists in masks. These were public and popular events. Their organizers and stakeholders might have had little to no interest in putting the law back into the hands of white people in view of the advancing Civil Rights Movement. Like the festivities around the auction block or white minstrel artists and producers, people knew they could make money in producing these events. Because anti-Blackness is the lowest common denominator for all whites, it was a surefire way to attract the masses. We should thus conceive of lynchings in their affective function for all white subjects involved: the organizers, the actual murderers, the distributors, the advertisers, and most importantly the audience. For all of them, lynchings provide an affect of racialized and anti-Black national identity. Baldwin does not select a random date for the picnic: this is the Fourth of July. At a lynching event, the white subject undergoes the profound experience of collectively shared anti-Black abjection in its utmost extreme, and thus generates a powerful subject-aeffect of belonging to his symbolic kin: a collectively experienced reinvigoration of modern, American subjectivity.

For 8-year-old Jesse, this lynching event, and specifically the moment of castration, is formative. It provides a way of identifying with the celebrated collective abjector, who is doing the work *for* the subject (a configurative dynamic that comes up in the performance of bawdry, as theorized by Freud and discussed in the chapter on humor below). Jesse strongly identifies with the man who castrated the hanging body, wishing "he had been that man," who approached the "hanging gleaming body, the most terrible and beautiful object he had ever seen till then." His father's friend provided agency (into which the young boy could project himself) in a whole procedure that fetishizes the Black body's genitalia. Jesse takes in how his mother was "more beautiful than he had ever seen her, and more strange" (247).

Soon, young Jesse has sexual sensations (the first?) on his own:

> The man with the knife took the [n-word]'s privates in his hand, one hand, still smiling, as though he were weighing them. In the cradle of the one white hand, the [n-word]'s privates seemed as remote as meat being weighed in the scales; but seemed heavier, too, much heavier, and Jesse felt his scrotum tighten; and huge, huge, much bigger than his father's, flaccid, hairless, the largest thing he had ever seen till then, and the blackest. The white hand stretched them, cradled them, caressed them. Then the dying man's eyes looked straight into Jesse's eyes – it could not have been more than a second, but it seemed longer

than a year. Then Jesse screamed as the knife flashed, first up, then down, cutting the dreadful thing away, and the blood came roaring down. (247–48)

As the continuation of anti-Black dehumanizing abjective violence, lynching is a space where we can observe what Hartman terms the terror of pleasure and pleasure of terror in a highly sexual form. Baldwin's story repeatedly illustrates how anti-Black abjection creates a subject-aeffect through desire and fear or disgust. Seeking to overcome erectile dysfunction, Jesse's adult mind – consciously or unconsciously – draws on abjective fantasies, of which the lynching event is only the last. His childhood memory of the lynching festivity is preceded by the imagination of a "Black girl [causing] a distant excitement in him, like a far-away light." However, "the excitement was more like pain; instead of forcing him to act, it made action impossible." For Jesse's libidinal economy, there is no differentiation between a "black piece" being "arrested" or "picked up:" "it came to the same thing" (229–30). The specific modality of dehumanization is not part of the issue; both fall into the same affective function of anti-Black abjection.

The indistinction between fear or desire is further marked when Jesse goes on to think about encounters with people racialized as Black in his function as a deputy sheriff: "He felt that he would like to [...] never again feel that filthy, kinky, greasy hair under his hand, never again watch those black breasts leap against the leaping cattle prod, never hear those moans again or watch that blood run down or the fat lips split or the sealed eyes struggle open" (230–31). The imagination of the Black girl does not have the desired effect, and Jesse's (unconscious) mind then brings him to recall a situation from work, in which he demonstratively beat up the "ring-leader" (232) of a group of Civil Rights activists to set an example. He shares his memory with his wife, but is unsure "whether she was listening or not" (thus in virtual solipsism):

> He was lying on the ground jerking and moaning, they had thrown him in a cell all by himself, and blood was coming out of his ears from where Big Jim C. and his boys had whipped him. Wouldn't you think they'd learn? I put the prod to him and he jerked some more and he kind of screamed – but he didn't have much voice left. (232)

In remembering, recounting, and re-performing this narrative, "he began to hurt all over with that peculiar excitement which refused to be relieved" (232). Again, Jesse's sexual energy, cathected in the beating of the Black body almost to death, does not differentiate between fear or desire. I am reminded of Kristeva's dictum that fear is an "abortive metaphor of want" (35). After beating the boy unconscious, Jesse "was shaking more than the boy had been shaking. He was glad that no one could see him. At the same time, he felt very close to a very peculiar, particular joy; something deep in him and deep in his memory was stirred, but whatever was in his memory eluded him" (233). Baldwin pushes the abjective force to its logical extreme: just as he is walking to the cell door, Jesse is addressed by the young Civil Rights leader he thought to be unconscious, and "[for] some reason, he grabbed his privates" (233). He then performs anti-Black abjection of an already apparently dead Black body. Because the boy is unlikely to talk back any more, and because there is no audience, this abjection is an entirely autopoietic, solipsistic phenomenon (much like recounting it to his tuned-out wife) that provides a subject-aeffect ending in involuntary sexual arousal.

> Now the boy looked as though he were dead. Jesse wanted to go over him and pick him up and pistol whip him until the boy's head burst open like a melon. He began to tremble with what he believed was rage, sweat, both cold and hot, raced down his body, the singing filled him as though it were a weird, uncontrollable monstrous howling rumbling from the depths of his own belly, he felt an icy fear rise in him and raise him up, and he shouted, he howled, "You lucky we *pump* some white blood into you every once in a while – your women! Here's what I got for all the black bitches in the world–!" Then he was, abruptly, almost too weak to stand; to his bewilderment, his horror, beneath his own fingers, he felt himself violently stiffen – with no warning at all. (235)

This passage eloquently crystallizes much of Baldwin's narrative as well as my argument so far: the cold and hot sweat, the singing Jesse sought to stop that suddenly comes from his own belly, the fear rising inside that raises him up. None of this is voluntary. There is "no warning at all" for these affects. Not only is it involuntary; it is also to some degree incomprehensible for the white subject personified by Jesse. His feelings and sensations are not for him to understand; "whatever was in his memory eluded him" (233), and he can talk only superficially about "what he believed was rage," and a generally "obscure comfort" that comes with certain songs he hears (235). Jesse, the white subject, does not understand the language of his libido, the grammar by which he lives.

In an example of pure and almost textbook anti-Black abjection, he then puts on an *imagined blackface*. Jesse – whom the reader has followed as he ponders various dehumanizing modes of gratuitous violence, both symbolic and utterly reality – ultimately overcomes his erectile dysfunction by imagining himself in terms of the very Blackness he abjects:

> He thought of the boy in the cell; he thought of the man in the fire; he thought of the knife and grabbed himself and stroked himself and a terrible sound, something between a high laugh and a howl, came out of him and dragged his sleeping wife up on one elbow. [...] He thought of the morning and grabbed her, laughing and crying, crying and laughing, and he whispered, as he stroked her, as he took her, "Come on, sugar, I'm going to do you like a [n-

word], come on, sugar, and love me just like you'd love a [n-word]." He thought of the morning, as he labored and she moaned, thought of the morning as he labored harder than he ever had before. (249)

The libidinal economy of the modern Western subject plays out through the sexual sphere of subject-aeffects, always grounded in the anti-Blackness that stimulates it. The white subject's simultaneous destruction of and desire for Blackness is central in this continual procedure, which is why "white people fuck on-screen to black music and not to their own," as Wilderson notes (Williams and Wilderson, par. 115). Wilderson makes the instructive point that while everyone knows *intuitively* that this is true, it remains *intellectually* difficult to grasp for many:

> I think everybody in this country and everywhere I've been in the world wraps their heads around it intuitively, which is why blackness is so energizing, whether it's negrophilia or negrophobia. The energizing capacity of blackness is just infinite because it's this locus of violence from which respite cannot even be theorized. (par. 118)

Blackness enables the white subject to experience itself in a way that feels true, real, and coherent – symbolically, libidinally, sexually. For the white subject, Blackness "authenticates the orgasm, a kind of pure jouissance" (par. 118).

"Now put on this noose:" Improvising the cultural repertoire
In its extreme configuration of fear/desire, lynching is a part of the US entertainment industry's repertoire and provides what Goddard and Wierzbicka call a cultural script. As developed and applied in the field of pragmatic linguistics, a "cultural script" is "a technique for articulating cultural norms, values, and practices in terms which are clear, precise, and accessible to cultural insiders and outsiders alike" (153). They write:

> Aside from the semantics of cultural key words, other kinds of linguistic evidence which can be particularly revealing of cultural norms and values include: common sayings and proverbs, frequent collocations, conversational routines and varieties of formulaic or semi-formulaic speech, discourse particles and interjections, and terms of address and reference – all highly "interactional" aspects of language. (153–54)

Although Goddard and Wierzbicka are primarily concerned with linguistics, we should not hesitate to read culture as text here and understand the explanatory power of a concept like a script for culturally specific signification. They also position the concept within a broader field of ethnopragmatics, which resonates with Sylvia Wynter's work. The use of cultural scripts in communication will also, by necessity, provide a sense of being "a good man or woman of one's kind" (Davis qtd. in Wynter, "Unsettling" 271). In the context of anti-Black lynching, we must reckon with the fact that, in mobilizing the "spectacular nature of black suffering" (Hartman, *Scenes* 22), lynchings engender, articulate, and make manifest a popular culture that already exists as a libidinally (politically, legally) shared ground, representing a norm rather than an exception. They are to the popular as common sayings are to language. Both share the performative *pars pro toto* logic of knowing/speaking, ensuring that the speakers know where they *belong* and providing them the sensation of a subject-aeffect. In addition, lynchings communicate to the outsiders (to Humanity) racialized as Black in "clear, precise, and accessible" terms (and experiences). Because the script of lynching is not so much part of but effectively built into the notion of the popular culture as we know it today, it can be intuitively activated anytime – be it in concrete action, representational arts, or displaced metaphor – with full white integrity:

> I did a show recently in which a guy initiated a scene with me. He was a white male. He initiated the scene by, "Ok, now put on this noose." And I was like "Whoa, in real life how would I respond to this?" And I was like "Hey, I don't think" – because he was my father – "Hey father, I don't think you're being really sensitive to my cultural background, so I am not gonna do this." It was a response in character but also like, "How would I as a human in this situation respond to this?" It's yes and...! But you can yes-and something without dooming it. This was furthering the scene to a point. I don't have to comply to this theme that I think is not nice. I personally don't invite friends and family to my improv shows. I invite them to sketch shows where I know what is about to be said but improv? No. Eight weeks of shows – I am not inviting anyone because I know what can happen on stage. And then this one time, when a friend came, this lynch thing happened. (Perkins, personal conversation)

I will return to this anecdote, but here I only want to point out that the symbolic violence on the improv stage in this scene is a central – if superficial – aspect of how improvisers ensure that improv maintains a white space. As white as actual lynching sites and the towns or cities in which they take place – as in Baldwin's "Going to Meet the Man," when 8-year-old Jesse, driving with his family to the lynching party, realizes that

> he had not seen a black face anymore for more than two days [...] there were no black faces on the road this morning, no black people anywhere. From the houses in which they lived, all along the road, no smoke curled, no life stirred – maybe one or two chickens were to be seen, that was all. There was no one at the windows, no one in the yard, no one sitting on the porches, and the doors were closed. [...] They passed the [n-word] church – dead-white, desolate, locked up; and the graveyard, where no one knelt or walked, and he saw no flowers. He wanted to ask, *Where are they? Where are they all?* But he did not dare. (243–44)

In an unsettling analogy with the improviser curious about the absence of Black "life" or the fact that the Black church has turned "dead-white," Jesse lacks the courage to ask this question; he knows somehow that the answer lies in the structure of the situation, in what can be anticipated. For Black people, improv is less a safe space than one in which danger openly lurks, because collective entertainment centers around the symbolic, real, and symbolically real dehumanization, killing, burning, and castration of Black-racialized individuals. Choosing or advising others to stay absent is different than allegedly being insufficiently "cosmopolitan," as Roger Bowen has it (qtd. in Sweet 40). Many will (intuitively) deny a connection between anti-Black lynching parties and an improv scene, but the anti-Black dynamics are the same, and the abysmal performance of a lynching reenactment (or the suggestion of one) only takes what is always already present in other forms of staged entertainment to its logical extreme.

Historicizing the popular III: minstrelsy

I now turn to a more widely-accepted understanding of the role of Blackness in popular culture. Mel Watkins's *On the real side* is an extensive history, an invaluable archival work, and a multifaceted approach to Blackness, comedy, and US popular entertainment. Despite our conceptual and axiomatic differences, his work is highly informative and provides ample points of reference for my project.[23] Watkins notes that, when Black performers entered the popular stage, "a

---

[23] In view of the above theorization of critiquing the modern West, Blackness has not *incidentally* become attached to these significations. Watkins's extensive work engages with the question of humor within Black communities and the assumptive ascriptions made by white people towards Black humor and Blackness in general. His study "traces and examines the social functions of two disparate strains of humor: the often distorted *outside* presentation in mainstream media (initially by non-Blacks) and the authentic *inside* development of humor in Black communities (from slave shanties and street corners to cabarets) as well as in folklore and Black literature, films, and race records" (41). In this endeavor, Watkins presents a grand effort to isolate African elements of humor on the plantation. He provides ample evidence to understand the ingenuity and intelligence with which slaves adopted social masks, using humor in communicative functions that would exceed the white masters. In so doing and by necessity, the enslaved had from the very first moment a clearer understanding of the social configuration – and humor may have been a way of coping. Watkins quotes Ellison: "We couldn't escape, so we developed a style of humor which recognized the basic artificiality, the irrationality, of the actual arrangement" (33). Watkins argues within the terms of influence: "it is the expressive manner of African-American humor that, second to music, has most influenced mainstream America's popular culture" (48–49). In this project, I am concerned with how Blackness is a structural necessity for the emergence of popular entertainment, rather than suggesting ways in which it *influenced*

distorted black spectre already dominated the stage" (123). He addresses how Blackness serves as the ground for all American US entertainment, easily overwriting all the stock characters of humor European immigrants brought from their home nations: "blackness was associated with humor almost from the outset [...] but it was not until the early 1800s that [Black characters] began to emerge as principle figures in America's comic lexicon." Other comic types, whether regional or of other ethnicities, were slowly replaced by "black-faced caricatures" (82–83). While several white actors and clowns imitated (their interpretation of) Blackness in the 1820s, Thomas Rice's blackfaced stage persona became the biggest success. Advertised as "Jim Crow Rice," he "became one of America's best-known comedians." Rice's and other shows ensured that "[b]y the 1830s, blackfaced white performers were one of the most popular attractions on the American stage." By the end of that decade,

> through refinement and more determined exploitation of the subject, blackfaced characterization had virtually eliminated all other ethnic or regional types. "Jim Crow," the unkempt ignorant plantation slave, displaced the backwoods or Frontier caricature; and "Zip Coon" or "Jim Dandy," the bombastic dandified city slicker, replaced the Yankee character as America's central comic figures. (Watkins 84–86)

The overriding of European characters by Blackness represents how Americans unified qua Blackness in their distinction from their European legacy. Though by no means unknown to modern Europe, the anti-Black abjective aesthetics of minstrelsy were mobilized to create a unique national self-understanding of Americanness via a generalized yet specifically (anti-)Black popular entertainment. To be a "good man or woman of one's [North American] kind" (Davis qtd. in Wynter, "Unsettling" 271) was not *connected* to or involved in or part of this culture – it *was* it.[24] When the white American subject imagines its social

---

it. How exactly this happened and what specific comedic strategies were used to what ends is not for me to discuss – mainly because the vector of this inquiry would throw me into the ethnographic trap. I would necessarily fall victim to the white "misinterpretation of slave behavior," that was, according to Watkins, the ground for Black humor to emerge.

24 Notably, theatrical entertainment at that time "included not only a full-length play but an assortment of variety acts" (Watkins 85). These shows were "consciously low-brow entertainment [that] emphasized the spectacular and the bizarre" in a "raucous, sensational, and often profane atmosphere of popular American entertainment during this period" (86). Because "for Northerners, in particular, blacks were still seen as curiosities," blackface performances of white dehumanizing imaginations soon became part and parcel of the entertainment per se. They were intricately bound up semantically with the experience of jouissance in collective, popular, anti-European self-assurance through laughter in the aesthetics of theatrical representation.

and political distinction from Europe, the ground and motorizing energy for this political negotiation is Blackness, and its libidinal and aesthetic mode is anti-Black abjection.

Here I want to consider the genesis narrative of the first fully-fledged blackface minstrel troupe, though I maintain some distance from Watkins's notion of accidentality and his primarily economic framework:

> As with many events in American cultural history, the establishment of the minstrel show as a separate form of entertainment was accidental. America had experienced a financial panic in 1837, and in the early 1840s, the nation still reeled near the brink of financial disaster. Unemployment was rampant, and even among variety performers, jobs were hard to find. Seeking a solution to their own financial problems, four out-of-work white performers met in a New York City hotel in 1842 – a year that some historians described as the "nadir" of the theatrical scene. The men [...] all had previous experience as blackface entertainers. [...] The idea [to form a troupe and concentrate exclusively on blackface mimicry], while certainly opportunistic, was no more than a pragmatic solution to their immediate problems. They were unaware that they had stumbled upon a notion that would transform American entertainment and firmly establish the image of blacks as happy-go-lucky plantation darkies and outrageously dressed, ignorant dandies in the entertainment media. (81)

Rather than engaging with the contingency of the historical situation or reckoning with the economic pressures these white men faced before they came to fame as the Virginia Minstrels from 1843 onwards, I highlight the ease with which they could decide to "concentrate exclusively on black mimicry" because anti-Black abjective discourse and affect were readily available to them. Because its success was already there, Blackness already *was* popular comedic culture in the form of the auction block spectacle, before "entertainment culture" existed as understood today. There were no political or ethical boundaries, and a white community of laughers could safely be counted on because Blackness already existed for the white public's enjoyment. Anti-Blackness provides the structural ground, the semantic field of psychosocial abjection, on which the US entertainment industry was and is based. Minstrelsy did not *accidentally happen to emerge* in the "nadir" of the theater scene but galvanized the anti-Black forces structuring popular culture that were always already there. As a nationalized form of entertainment, "America's popular culture signature piece" was much more than lowbrow entertainment (Kopano 5). Quite the contrary, in their enjoyment of minstrelsy, the working class could feel aligned with the higher levels of society on the grounds of their shared skin color. By 1844, "this new entertainment genre had so swept the nation, that the Serenaders were invited to the White House to perform for the 'Especial amusement of the President of the United States'" (Watkins 88). This form of popular culture did not merely provide meaning to the notion of a white working-class (or the grounds for its creation in Eric

Lott's sense), but facilitated national cohesion. Minstrelsy posited a vital, democratic American man, invigorated by Blackness/blackface and distinct from the degenerate lifeless figures of Old Europe. (The same logic of distinction from Old Europe remains visible in the celebration of improv as an All-American art form as will be discussed later.)

Given its cohesive power, Watkins perceives minstrelsy in a period of nationalist crisis as a racial contact zone – a space of potential sociopolitical negotiation, which debatable position Eric Lott also holds. Watkins argues that during the superficial "rift" that was abolition, minstrelsy as national popular culture mobilized the essence, the affective ground of its success, and turned towards the overt abjective dehumanization of Blackness that had always libidinally structured it:

> Confronted with a choice of preserving the Union or supporting black Emancipation, [minstrel acts] soon eliminated all but the most servile and disparaging images of blacks from their shows. From about 1853 to the Civil War, then, nearly all vestiges of black humanity were excised from minstrel performances. During this period the portrait of the plantation was made even more idyllic, and the stereotype of black males as childlike, shiftless, irresponsible dolts was heightened. Freed blacks, in particular, came under pointed attack. They were invariably pictured as inept, hopelessly inadequate souls, who longed for the guidance of white men and the security of the 'ole plantation,' or, perhaps worse, near-bestial reprobates who, after disastrous consequences, foolishly took on 'white' airs and lusted after white women. The comic, degrading image of blacks had almost reached its peak. America's most popular entertainment form had become a forum in which white performers posing as blacks actively lobbied for the continuation of slavery by presenting degrading, consciously distorted comic stereotypes intended to 'prove' that slavery and black subordination were justified, or, even more insidiously, to demonstrate that blacks actually preferred serfdom. (94–95)

Preceded by the auction block and followed by popular lynching events, minstrelsy was a formative genre in the creation of popular culture, going far beyond the notions of making a white working class that Lott suggests. It riveted Blackness into the national unconscious, generating a space that allows white people not only to use, apply, and capitalize upon it, but also to *relish*, savor, and consume it. Even Watkins, in view of his recurring reference to historical contingency, is quite clear and mobilizes the trope of public property:

> Of course, the Sambo stereotype began long before minstrelsy. [...] But the popularity of the minstrel shows heightened its acceptance and riveted it into the national consciousness. By the 1880s, that image had become public *property* [...] Americans who detested flesh-and-blood blacks relished the minstrelsy-inspired caricatures that flooded the country. Minstrelsy made Sambo as American as apple pie. (102–03; emphasis mine)

However, minstrelsy was more than "the heart of 19th-century show business" (Tosches 11). It also set up a libidinal space in which white modern subjects could fantasize about the transcendence of what they perceived as the boundaries of their sociopolitical existence, their reduction to rationalism as determined in their ratiocentric descriptive statement. Burnt cork on the face was a vehicle for circumventing the restrictions of a self-imposed *civilized* (as opposed to *natural*) self-image, both for professional and amateur performers and for their audiences:

> Whether literally as a performer, or figuratively as an observer, "the white man who put on the black mask modeled himself after [...] a black man of lust and passion and natural freedom." He thereby not only indulged the desire to escape the binds of "civilized" behavior but also affirmed his superiority. (Watkins 100)

Blackness here functions as a portal for white flights of performative excellence, that is, being *very* good men of their kind, which white subjects (felt they) were unable to do without the comic cork. Watkins shows how the jouissance experienced in the audience was most likely shared by the actors on stage: "The cloak of blackness apparently allowed them to cast most of their own inhibitions to the wind, thereby heightening the excitement and frenetic pace of their performance [with] exuberance and vitality" (87). Anti-Blackness created clowns and provided the somaesthetic transcendence of the self through a culturally specific mask. Blackface also paved the way for an amateur theater scene, allowing anybody to perform per the egalitarian and democratic ideals of the republic. The abjective jouissance of partaking in blackface performance lured actors onto the amateur stage. By the turn of the century, "'every city, town, and rural community had amateur minstrel groups'" (Boskin qtd. in Watkins 99). In 1930, Carl Wittke, "an unabashed minstrel enthusiast," lamented the disappearance of the American minstrel show "except as a vehicle for amateurs" (qtd. in Watkins 98). Anti-Black abjection articulated through blackface was an easy way for white subjects to become amateur actors.

In our engagement with popular cultural practices, we must keep in mind that however universal or generic popular culture may seem, it has always been "uniquely African-American in origin, conception, and inspiration" in all its styles and modes (Tate 2). Like the early traders of enslaved Africans and plantation owners, the players in the field of popular culture would violently acquire "black bodies [that] would be responsible for providing the labor and natural resources that would propel all of the Western powerhouses to their global supremacy" (Kopano 4):

> Everything about African-Americans – their bodies, dances, songs, dialects, passions, worldview, and so on, real and especially as imagined by whites (and, eventually, imagined by other blacks, too) – became the base material for the popular entertainments that matured into mass culture. The black condition became a canvas for projecting all that the whiteness ideal sublimated; black cultural output became the paint. (L. Wynter 22–23)

In view of the above discussion of the auction block, lynching parties, and minstrelsy, it must be conceded that there is an overt and not particularly subtle continuum of dehumanizing anti-Blackness in the making of the popular and its culture as such. Minstrelsy is not a distinct phase of popular culture, but its ground. Minstrelsy provides the occasion, the aesthetics, the content, and everything else to modern US theatrical entertainment. It is the brick and mortar of a popular wall that may be painted or scribbled on, but will always function as the border of whiteness. Whenever white subjects seek to reassert ourselves through the popular, even in imagined opposition to it, whenever we feel drawn to (consume) popular culture, we are drawn to (consume) Blackness. When we are active in popular culture, we are putting on a "cloak of Blackness" in one way or another (Watkins 87). And it does not stop here. The anti-Black fabric of popular culture predetermines the way it has traditionally been theorized, as I will consider in the next section.

Theorizing the anti-Black popular

Current academic interest in popular culture tends to meander between two positions: on the one hand, a romanticized fascination with its (assumed) subversive potential that celebrates the possibilities of more or less radical oppositionality (whether aesthetically or politically framed, temporarily realized or imagined utopias); and on the other, as the capitalist top-to-bottom infiltration of hegemonic knowledges into an assumed political unconsciousness of the people (whom exactly?) via a *panis et circenses*-cultural industry (and subsequent differentiation between high and low art), most prominently developed in the work of Adorno and the Frankfurt School. These positions resonate with the simultaneity of fear and desire, terror and pleasure to which the anti-Black ground of popular culture gives rise. Metatheoretically, it is not farfetched to interpret *both* as libidinally motivated in order to generate a subject-aeffect. Consider Jackson's assertion that the Blackness of the field drives scholarly endeavor:

> [I]t could be conjectured that one of the main reasons "popular culture" has become a category for inquiry has been the enormous success of hip-hop culture and its component performance domains, rapping, graffiti writing, break dancing, emceeing, and deejaying. A

> huge intelligentsia – scholars, television and newspaper journalists, museum curators, a wide range of artists, hip-hop magazine cultural critics, and filmmakers – has been seduced by hip-hop's growth and vitality. Hip-hop has played a role in establishing the field of black mass cultural studies. (23)

Engaging with a phenomenon like hip-hop culture on academic turf does not happen outside the affective register. Quite the contrary, when performed by a white scholar, dealing with Black cultural material is in itself the performance of anti-Black abjection that ultimately reasserts the white scholar as subject and obliterates the object in a kind of culturally specific logorrhea. While this critique can certainly be directed against the present project, it is especially true if the academic gesture stands in the ethnographic tradition, using its methodology or applying its axiomatic logic – which also involves the mobilization of concepts like authenticity and appropriation. This is especially difficult to grasp for those with a positive, defensive, or apologetic view of popular culture, who assume that, with the help of popular cultural material and the analysis thereof, the differentiation between low and high art can be overcome.

Shusterman's romance

In "The Fine Art of Rap," Richard Shusterman – who "likes the music" and thus has "a personal stake in defending its aesthetic legitimacy" (201) – addresses the qualities of rap, arguing that they satisfy "the most crucial conventional criteria for aesthetic legitimacy" (202). In his discussion, he applies the postmodern toolbox and draws on the modern modalities of intellectual autonomy. As well-meaning as he may be, and as progressive as his approach may appear, it pretends to (temporarily) lend subjective capacity to Blackness qua white scholarly authority, which must always remain at the mercy of the master-scholar. Additionally, it assumes that an *aesthetic-scholarly legitimization within white academic institutionality* is in some way desirable for a cultural scene that has no stake in it or anybody else. Both aspects position the argument in a feigned register of negotiation that is in actuality foreclosed to Shusterman's object itself. His treatment of rock music in "Animadversions on the Critique of Popular Art," which pits the somatic aesthetics of rock music against intellectual highbrow-treatment, is a similarly racial celebration of Blackness-as-popular culture. He writes:

> Rock songs are typically enjoyed through moving, dancing, and singing along with the music, often with such vigorous effort that we break into a sweat and eventually exhaust ourselves. And such efforts, as Dewey realized, involve overcoming such resistances as "embarrassment, fear, awkwardness, self-consciousness, [and] lack of vitality" (Dewey

> 1987 [1934]: 162) [...] The term "funky," used to characterize and commend many rock songs, derives from an Africa word meaning "positive sweat" and is expressive on an African aesthetic of vigorously active and communally impassioned engagement, as opposed to dispassionate, analytical remoteness. The much more energetic and kinesthetic aesthetic response evoked by rock exposes the fundamental passivity underlying our established appreciation of high art. (111)

Shusterman deploys the term "funk" from jazz parlance in this argument about (the largely white phenomenon of) rock music.[25] He makes violently homogenizing assumptions about the unified aesthetic of an entire continent, pitching "passion, sweat and energy" against a European "dispassionate analytical remoteness" and thus vigorously remains within the axiomatic logic originating in the Platonic postulate, transumed in the theocentric statement of Heavens and Earth, and turned into the racialized dichotomy between High Reason and irrationality, between intellect and sensuality. In Shusterman's work, we can observe that such a positive affirmation of popular culture's *sensuality* is only possible qua racialized reference to the specific Blackness of popular cultural production. This is not his *fault* – the register of morality is not helpful here – but is instead a discursive predisposition of the system in which popular culture exists *as Blackness* without being called so. If we want to argue in favor of popular culture, we have no alternative to applying the racial register of its Black physicality, baseness, earthiness, affect.[26] We can further substantiate this position by considering how other (unfavorable) treatments of popular culture draw on the same register: Those who despise it do so on the same argumentational ground. Where Shusterman cherishes the danceability of rock music as an almost transcendental experience, this is also at the center of Adorno's infamous jazz critique. Where Shusterman's theory is structured by desire for Blackness-as-Popular Culture, fear speaks through Adorno's. And both perform a dazzling dance on the (non-existent) thin line between the witness and the spectator.

---

25 I want to note not only that rock music is predominantly a white, male phenomenon, but that its entrance into popular culture is linked to processes of Black cultural production's obliteration, as Waksman and others recognize.

26 I hope it has become clear that this discursive dimension is not superficial or detached from the real world. This entire chapter has argued that popular culture's discursive dimension and its racialized rhetorical repertoire are entirely grounded in traceable historicity, numerous real-world performances, and their contemporary variants. This is not a matter of *interpreting* the popular, but of reframing it in its historicity and cultural specificity.

Adorno's Querfront

Adorno's treatment of jazz has been much discussed, attacked, and defended. However, to get a grip on the incommunicability between scholar and object and the surprising consequences that follow from his line of argument, we must not only take his theoretical context very seriously but also reckon with the racial dimension and libidinal terminology he applies. Adorno's racism cannot be excused; we can neither bracket instances of overtly racial language out of an otherwise valuable jazz critique nor cut that critique out of an otherwise valuable critical framework. The racialized components of his analysis are not accidental; his allegedly objective musicological arguments only operate when substantiated and substantialized by anti-Blackness. If, as I have argued, popular culture is made up of Blackness in its modes, material, and the affective structures it mobilizes, then what Adorno has to say *within* the anti-Black register will have some bearing *on* this register. I therefore read his racism and his anti-Black rhetorics as symptoms of the foundational structure and libidinal investment in his critique of popular or mass culture and the cultural industry at large, which was a project designed to re-install modern subjectivity as such – the Human of Humanism – after the atrocities of World War II.

After decades of superficial debates between musicological justifications ("It is true that jazz musicians did not *invent* novel harmonies!"), blunt accusations that Adorno is ignorant of the "social" dimensions of jazz ("It is important for *the* Black community!"), aggressive positions on moral grounds ("White people must (not) be allowed to speak negatively about Black cultural production!"), and defenses of jazz on equally racial grounds ("Adorno might not like it, but I appreciate Black performers expressing their wildness!"), in recent years the engagement with this complex has matured. Suggesting that "all the cards in this game have been played," Eric Oberle regards Adorno's jazz critique as a "wound" in the latter's critical oeuvre. Oberle reads this "wound" symptomatically, not unlike the punctum methodology I apply in this project, given that "the wound of jazz is defined by problems of race, culture, identity, violence, and discrimination." Oberle positions Adorno "among the twentieth century's pioneering critics and analysts of racism and cultural bigotry," and finds that his treatments of jazz stand in paradoxical relation to the rest of his work (365). Oberle has "no doubt that the jazz question must be taken seriously as a limitation that reflects much about Adorno, both biographically and as a philosopher, sociologist, and cultural critic" (365). In terms of a more mature engagement with Adorno's criticism, Oberle suggests relating what he refers to as "conceptual issues" to "the problems they were hoping to solve, and to how those ideas necessarily struck against their limits – limits with which the individual thinker had difficulty, and which he could do little else but to internalize" (365).[27] However, he does not provide the ultimate guide on how to read Adorno's relationship with jazz constructively. There are, in fact, some cards yet to play (and presumably several more not covered here). Fumi Okiji's *Jazz as Critique: Adorno and Black Expression Revisited* is the most instructive discussion of the subject. The author brings to the table the musicological expertise of a professional jazz performer, the scholarly acuteness of the critical theorist, and the competency to talk about the functions of Blackness meaningfully. No more does she ignore the racial rhetoric of Adorno's critique but takes on a different argumentational direction by suggesting that Black existence and the cultural production emerging from it are always already "critical" of white modernity:

> Blackness may well be a thing not yet known, as Fred Moten tells us, and it is unclear how the world could ever know it without internal collapse. But black life *is* lived, and particularly where it comes up against its appropriated and sanctioned mainstream images and uses, where it misshapes the categorical smoothness of race, it provides valuable insight. In its contradictory subjecthood – human enough for governance but too black for admittance in the "household of humanity"– such life rhymes with what Adorno understands as the double character of radical art, rejecting what it is unable to rid itself of through critical immersion. [...] What is suggested here [...] is that black expressive work cannot but help shed light on black life's (im)possibilities. (4)

According to Okiji, jazz is not only "capable of reflecting critically on the contradictions from which it arises [but] it is compelled to do so" (5). Throughout *Jazz as Critique*, and drawing from a wide range of Black theorists (most centrally Du Bois), she demonstrates convincingly that jazz can indeed be argued to perform Adorno's aesthetic demands for subjective and aesthetic resistance – precisely because of its Blackness. This argument assumes Adorno's ignorance both of "the principles of structuration in jazz work" and of "black sociohistory" (12). Okiji sets out to order and qualify the discussion. Even though her approach is highly instructive, I do not go down this path because a) I see no need to legitimize either Adorno or jazz, b) I wish to avoid staying within the register of negotiation and resistance as developed *by* Adorno, and c) I cannot add to what Okiji says about jazz practice. For the remainder of this chapter, I will work with her insight occasionally but not instructively.

---

[27] Oberle theorizes of twentieth-century "identity" and a "wounded political subjectivity of the modern era" (357). As I am not interested in reading Adorno's critical theory at large, especially in view of the white, Human disposition of wounded-ness of a universalized subject, I will not engage with his text in more detail.

Adorno did not mistake white commercialized swing for "real" jazz, but was aware of the development of bebop and "oppositional groups" ("Fashion" 122).[28] However, for him, all jazz is primarily a "type of dance music" (45), "musically completely banal and conventional" (67), and all of its variants are governed by the same aesthetic principles, which never fundamentally break the "harmonic-melodic convention of traditional dance music" ("On Jazz" 45). Syncopation is presented as jazz's rhythmic structuring principle. Accordingly, we can deduce that when Adorno speaks about jazz, he also means rock and roll and any other kind of *popular music* before and after. He inflates jazz to encompass all popular music.[29] He follows the same logic when addressing the device of vibrato:

> [Jazz's] vital component is the vibrato which causes a tone which is rigid and objective to tremble as if on its own; it ascribes to it subjective emotions without this being allowed to interrupt the fixedness of the basic sound-pattern, just as the syncopation is not allowed to interrupt the basic meter. (46)

He later suggests that the "jazz-sound itself [...] is determined by the possibility of letting the rigid vibrate, or more generally, by the opportunity to produce interferences between the rigid and the excessive" (46–47). For Adorno, the vibrato is a temporary fiction of subjective, individualized resistance against the confinements of a universalized modern existence, because it is always bound to the basic sound-pattern – or so the argument goes. His critique of syncopation is more elaborate. He argues that syncopation is a diversion from the capitalist mechanisms of the cultural industry, veiling the machinic, rigid, capitalist, military, fascist pre-dominance of the fundamental beat rather than subversively opposing it:

> Syncopation is its rhythmic principle. It occurs in a variety of modifications, in addition to its elemental form (as the "cake walk," jazz's precursor, uses it), modifications which remain constantly permeated by this elemental form. The most commonly used modifications are the displacement of the basic rhythm through deletions (the Charleston) or slurring

---

28 He writes: "The wild antics of the first jazz bands from the South, New Orleans above all, and those from Chicago, have been toned down with the growth of commercialization and of the audience, and continued scholarly efforts to recover some of this original animation, whether called 'swing' or 'bebop,' inexorably succumb to commercial requirements and quickly lose their sting" ("Fashion" 119).

29 However, what appears a diffuse generalization is significantly valid in the sense that all US popular culture – even the notion of an American "popular" as such – can indeed be understood as Black, and Adorno's argument abjectively veers toward the specific Blackness of jazz.

(Ragtime);[30] "false" rhythm, more or less a treatment of common time as a result of three & three & two eight-notes, with the accent always on the first note of the group which stands out as a "false" beat (Scheintakt) from the principle rhythm; finally the "break," a cadence which is similar to an improvisation mostly at the end of the middle part two beats before the repetition of the principle part of the refrain. In all of these syncopations, which occasionally in virtuoso pieces yield an extraordinary complexity, the fundamental beat is rigorously maintained; it is marked over and over again by the bass drum. ("On Jazz" 45–46)

Because of these rhythmic and harmonic restrictions and the presumed fundamental acceptance of the symmetry ideal (metrical or harmonic), aesthetic innovation as much as subjective resistance is always already prefigured by an axiomatic definition of the expressible.

It follows for Adorno that true improvisation (axiomatically equated with innovation) is impossible in jazz, as it is pre-programmed and bound to patterns of modest range. In his view, the jazz musician is too restricted ever to transcend the "perennial sameness" of jazz, which does not allow for innovation but only exalts in "well-defined tricks, formulas and clichés" ("Fashion" 122). In "On Jazz" he writes: "Even the much-invoked improvisations, the 'hot' passages and breaks, are merely ornamental in their significance, and never part of the overall construction or determinant of the form" (53). Improvisations are always reducible to "the more or less feeble rehashing of basic formulas in which the schema shines through at every moment," which is apparently commonly understood because

> any precocious American teenager knows that the routine today scarcely leaves any room for improvisation, and that what appears as spontaneity is in fact carefully planned out in advance with machinelike precision. But even where there is real improvisation, in oppositional groups which perhaps even today still indulge in such things out of sheer pleasure, the sole material remains popular songs. ("Fashion" 122)

---

30 This historical trajectory, whether culturally or musicologically valid, demonstrates that when talking jazz, Adorno talks Blackness – either by way of aesthetic trajectory or in the dismissal of US popular culture, that is, of white people imitating Black cultural practice. His aesthetic line of argument traces jazz back to previous Black popular cultural practices. Watkins writes: "By the 1910s, the cakewalk, "coon" songs, and ragtime music – all with inspiration or origin in black communities – had begun dominating America's popular entertainment." The influence was such that the "acceptance of the cakewalk by white Americans marked a major change in manners" and also "subsequent dance rages such as the Turkey Trot, Charleston, and Black Bottom (all with black origins) spurred the shift in white America's dance habits" (145). He further argued that "as whites increasingly copied black dance steps, black music with its emphasis on rhythm and syncopation also increased in popularity [...] making danceability the key to a song's potential." Watkin then provides much insight into the development of ragtime out of both "coon" and cakewalk songs (143–48).

Okiji points out that the notion of individual resistance and individualism, as it refers to a bourgeois universalized self, says little about Black performance. She asserts that the

> individual holds a problematic but central position in jazz narratives [...] Its use has assisted the desire to bring jazz closer to the model provided by Western European concert music and the singularity of the composer and her or his composition. It is an abstraction that leads to the fetishization of the solo as the essence of jazz work. (7)

Individualism in jazz discourse manifests the compositional paradigm in which an artist creates an artistic piece, a world, out of their individual potency. This potency is based on the impotence or the discursive incapacity to be a subject in the white modern matrical episteme for those racialized as Black. Whenever individualist compositionalism is the paradigm for analysis, the grammar of that analysis has no hold on the performance reality of Blackness on or off stage. Viewing jazz "through a lens that sees composition as the predominant site of artistry," idealizing autonomous and individualistic identities as (fictionally) manifest rationalized yet romanticized composition, and endowing them with the potential for political resistance, Adorno cannot but look down on what he interprets as "counterfeit identities" in jazz (Okiji 19). Okiji's analysis is central for understanding Adorno's argument through the lens of this project. Not only does Adorno activate the romantic idealism of high, bourgeois art (the Ancient stars, the Christian heavens, modern rational idealism) as opposed to the base, non-homogenous nadir of the earth and its popular cultural production.[31]

Because of this active abjective ignorance, Adorno's modern bourgeois individualism and the compositional paradigm in which he formulates his critique both fall in line with Hitler Germany's propaganda. The first partial prohibition of jazz on German radio dates from 1933; the announcement by radio *Berliner Funkstunde* already invokes the idea of "degenerate art" and mobilizes hypersexualized negrophobia as well as the danceability of Black music as opposed to its aesthetics, which is linked to a vague notion of "German feeling":

> The Berliner Funkstunde banishes all questionable dance music, described as [n-word] music by the healthy common sense of the people, in which a salacious rhythm prevails and the melody is raped. The Funkstunde will continue to cultivate modern dance

---

[31] Interestingly, in his defense of "improvisation," white clarinetist Ted Gioia draws immediately from the matrical binary in naming his publication *The Imperfect Art*.

music, as long as it is not inartistic in its musical elements or violates German sentiment. (qtd. in Fark 165, my translation)[32]

These ideas feature in Adorno's more elaborate jazz critique. His 1933 "Abschied vom Jazz" ("Farewell to Jazz") embraces the Nazi prohibition of jazz and an explicitly racial argument legitimized in the allegedly neutral terms of aesthetic quality judgment:

> The regulation only confirmed by drastic verdict what has long since been decided on factual grounds: that jazz music itself has come to an end. Because no matter what you want to understand by white and [n-word] jazz, there is nothing to save here. Jazz has been hollowed out by its own stupidity. The regulation does not eliminate the musical influence of the [n-word] race on the northern one, and neither Cultural Bolshevism, but a piece of bad art. (qtd. in Fark 166, my translation)[33]

Adorno's logic actively partakes in Nazi cultural politics fueled by the notion of [n-word] jazz "raping melody," ultimately leading to its complete prohibition on the grounds that "[n-word] jazz" is corrosive and "destroys the foundation of our entire culture," as Fark quotes *Reichssendeleiter* Eugen Hadamowski (166).

On the grounds of both Hitler Germany's and the left-liberal anti-Black thought and affective structure, we can see how Adorno and the fascist propaganda machine have a common aim, a shared vector of self-making abjection. From a contemporary perspective, Adorno enters an uncanny alliance with fascist political action (resulting from propaganda about the racial sanitization of society that sought to eradicate the cultural influence of the "[n-word] race on the northern one"), veiling this abjectively powerful vector in an argument about aesthetic judgment. This constellation only appears more paradoxically absurd when Adorno uses jazz as a musical genre in which individual resistance is feigned to *serve* a fascist regime, based on what he reads as jazz's primary

---

32 "Die Berliner Funkstunde verbannt alle fragwürdige, vom gesunden Volksempfinden als '[n-word]musik bezeichnete Tanzmusik, in der ein aufreizender Rhythmus vorherrscht und die Melodik vergewaltigt wird. Die Funkstunde wird aber auch weiterhin moderne Tanzmusik pflegen, soweit sie in ihren musikalischen Elementen nicht unkünstlerisch ist oder deutsches Empfinden verletzt."

33 "Die Verordnung [...] hat [...] nur durchs drastische Verdikt bestätigt, was sachlich längst entschieden ist: das Ende der Jazzmusik selber. Denn gleichgültig, was man unter weißen und unter [n-word]jazz verstehen will, hier gibt es nichts zu retten [...] Was [...] den Jazz aushöhlte, ist eine eigene Stupidität. Mit ihm wird nicht der musikalische Einfluss der [N-wort]rasse auf die nördliche ausgemerzt; auch kein Kulturbolschewismus, sondern ein Stück schlechtes Kunstgewerbe."

rhythmic structure and instrumentation and making this tendency toward abstract fascism his central culturo-musicological criticism:

> The effectiveness of the principle of march music in jazz is evident. The basic rhythm of the continuo and the bass drum is completely in sync with march rhythm, and, since the introduction of six-eight time, jazz could be transformed effortlessly into a march. The connection here is historically grounded; one of the horns used in jazz is called the Sousaphone, after the march composer. Not only the saxophone has been borrowed from the military orchestra; the entire arrangement of the jazz orchestra, in terms of the melodic, bass, "obligatory," and mere filler instruments, is identical to that of a military band. Thus jazz can be easily adapted for use by fascism. ("On Jazz" 61)

Adorno's argument aligns with historical fascism in an anti-Black affective thrust in 1933, using the same abjective procedure to generate a metaphor of jazz as a model for the workings of fascism after World War II. Unsurprisingly, comparable *querfront* alignments occur in contemporary debates on the artistic value and moral judgment of hip-hop culture, especially gangsta rap. Here too left-liberal criticism (journalistic or academic) and racist common sense abject Black cultural production, reinforcing each other on common ground. However, rather than falling for the moral argument, I engage with Adorno's elaborations on their own turf. Given Germany's history, Adorno set out to devise a theory of resistance after Auschwitz, seeking to generate an aesthetic position from which actual resistance would be possible, where political action and negotiability as such were thinkable. Even though he might not have used terms like "betterment," "progress," or "improvement," his argument is necessarily positioned on the plane of political negotiation and linear development. He therefore partakes in what Warren would call a politics of hope within the sphere of governmentality, again facilitated by anti-Blackness.

"Feigned subjectivity"

For Adorno, jazz symbolizes the relationship between economy and culture. The industry of commodification knows art or artistic expression only as amusement, and the capitalist system generates dynamics that commodify resistance. These aesthetic arguments cannot be distinguished from his political project of restoring the modern subject after the atrocities it caused in World War II. For Adorno, such moral reassurance is possible via art as resistance. Unlike whatever happens in the lofty realms of high art, he argues, commodified cultural expression like jazz can never self-empower the subject. Political opposition can never be achieved because, structurally, it does not come from a place that offers the capacity for such actual opposition. On this larger scale, Adorno suggests that jazz performs "a subjectivity which revolts against a collective power which it itself *is*; for this reason, its revolt seems ridiculous and is beaten down by the drum just as syncopation is by the beat" ("On Jazz" 68). The brief English-language summary at the end of the original publication of "Über Jazz," under the pseudonym Hektor Rottweiler, is succinct: "The antagonistic character of jazz is expressed by the formula that the 'subject of jazz' permits itself to be annihilated by society in order to feel itself endorsed and vindicated by society" (258). In the "Oxforder Nachträge," an addendum to "Über Jazz" published in *Gesammelte Werke*, Adorno writes: "What is crucial about the jazz-subject is, that despite its individual character, *it does not own itself at all*" ("Über Jazz" 258, emphasis mine).

Those who have engaged with the legacies of slavery cannot but hear the anti-Blackness of such a formulation. But we may also ask: exactly who is meant by the "jazz-subject?" In a footnote, Okiji addresses this "real confusion about what subject is being referred to at various points in 'Über Jazz,'" which leads some to think that the subject in question is the bourgeois consumer of jazz, and others to think of the (Black) musician as that subject. Okiji interprets this as both a theoretical and rhetorical device, pointing to the "implications when Adorno talks of castration, clowns, and slaves, especially when we're told of 'oppressed people' being particularly well-adapted for jazz and life und monopolized capitalism" (101). Even though Adorno regularly makes the point that jazz does not represent some sort of originary Blackness, it would be a compulsively contrived position to delink Blackness from the argument, especially in his assumption of jazz's natural servility, as formulated for example in "On Jazz":

> Psychologically, the primal structure of jazz (Ur-Jazz) may most closely suggest the spontaneous singing of servant girls. Society has drawn its vital music – provided that it has not been made to order from the very beginning – not from the wild, but from the domesticated body in bondage. The sadomasochistic elements in jazz could be clearly connected to this. ("On Jazz" 53)

Moreover, Adorno emphasizes that it is not external market forces or sociopolitical structures but jazz and its practitioners themselves that are to be blamed; their submission is part of their being-in-the-world. Jazz is by its aesthetic modality authority-bound, in the oedipal Freudian sense:

> However little doubt there can be regarding the African elements in jazz, it is no less certain that everything unruly in it was from the very beginning integrated into a strict scheme, that its rebellious gestures are accompanied by the tendency to blind obeisance, much like the sadomasochistic type described by analytic psychology, the person who chafes

against the father-figure while secretly admiring him, who seeks to emulate him and in turn derives enjoyment from the subordination he overtly detests [...] It is not as though scurrilous businessmen have corrupted the voice of nature by attacking it from without; jazz takes care of this all by itself. ("Fashion" 121)

In view of jazz's Blackness, Adorno moves to a tendentious, bawdry castration of Blackness to showcase his own enlightened, high art, European potency:

The syncopation [in jazz] is not, like its counter-part, that of Beethoven, the expression of an accumulated subjective force which directed itself against authority until it had produced a new law out of itself. It is purposeless; it leads nowhere and is arbitrarily withdrawn by an undialectical, mathematical incorporation into the beat. It is plainly a "coming-too- early," just as anxiety leads to premature orgasm, just as impotence expresses itself through premature and incomplete orgasm. ("On Jazz" 66)

In "Perennial Fashion" he goes even further:

The aim of jazz is the mechanical reproduction of a regressive moment, a castration symbolism. "Give up your masculinity, let yourself be castrated," the eunuch-like sound of the jazz band both mocks and proclaims, "and you will be rewarded, accepted into a fraternity which shares the mystery of impotence with you, a mystery revealed at the moment of the initiation rite." (128–29)

In this line of thought, it follows that the jazz subject exists only as "the amalgam of a destroyed subjectivity and of the social power which produces it, eliminates it, and objectifies it through this elimination" ("On Jazz" 67). When it does express itself, it says "I am nothing, I am filth, no matter what they do to me, it serves me right" ("Fashion" 131).

But who ever said that one could derive the psychological pathology of a subject from a rhythmic structure? And why blow that fantasy up and map it metonymically onto collective cultural spaces, abjected or not? Most of Adorno's interpreters, apologists and critics alike, seem to adopt this far-fetched and highly consequential assumption, which has some grotesque excrescences. For example, Adorno states that "[j]azz and pogrom belong together" implying that the jazz-subject is either naturally fit for being pogromed or a stupid enough exception that consciously and voluntarily engages in jazz one way or another ("Über Jazz," *Gesammelte Werke* 101, my translation). Some lines of logic become almost facetious, as when Adorno treats his conceptual analogy as causality in the following lofty statements:

I clearly remember how shocked I was when I first read the word "jazz." It would be plausible that it comes from the German word "Hatz," sketching the pursuit of a slow dog by bloodhounds. The typeface appears to contain the same castration threat as that of the jazz orchestra with the grand piano's open lid [...] The name of jazz's final predecessor also belongs in the same context: ragtime [...] The "ragging of time" by the syncope is ambivalent. It is an expression of an oppositional feigned subjectivity revolting against the measure of time, and simultaneously that of a regression mapped out by the objective instance. ("Über Jazz," *Gesammelte Werke* 102, my translation)[34]

Many elements of this line of argument are worth pointing out: the repetition of an alleged castration threat that speaks to culture and society through a musicological argument; the assumptive positing of an aesthetic trajectory in which the predecessor, like jazz, is presented as regressive; the etymological coercion and grotesque (non!-)causality that Adorno performs to satisfy the logic of the argument. The last point crystallizes the argumentational performance of his jazz critique writ large. Even though etymological arguments of this kind always need to be considered with extreme caution, by relating the German "Hatz" and the English "jazz," Adorno's argument becomes idiosyncratic rather than providing the structural analysis he seeks. The logic (metaphor, simile, analogy, causality?) only operates at all when "jazz" is pronounced with a particularly strong German accent. It is safe to say that strong German accents have no causal bearings on the history of jazz, its aesthetics, social formations, actual performances, or anything else related to it. Yet within his argument, it simply fits: Adorno takes an English language term and articulates it in his own language to make it fit his own theorization and the axiomatic, a priori ground specific to it, actively ignoring what would be potential ("plausible") limitations – also specific to it. What appears to be a grotesque distortion of the facts and a violent coercion of reality turns out to be a succinct example of the solipsism of modern white critical thought. Regardless of one's position on popular culture and the Blackness by which it exists, it cannot be talked about without overt or covert, conscious or unconscious reference to this very Blackness. The racial absurdities and violent de-humanizations (un-subjectivations, de-masculations) that constitute Adorno's logic right up to the preposterous Hatz-jazz-etymology are not *wounds* or *mishaps* in the larger argument. They cannot (and do not need to be) excused on the grounds that Adorno had no real (or not much) knowledge about jazz,

---

34 "Ich erinnere mich deutlich, daß ich erschrak, als ich das Wort Jazz zum ersten Male las. Plausibel wäre, daß es vom deutschen Wort Hatz kommt und die Verfolgung eines Langsameren durch Bluthunde entwirft. Jedenfalls scheint das Schriftbild die gleiche Kastrationsbedrohung zu enthalten, die das des Jazzorchesters mit dem aufgesperrten Flügeldeckel darstellt [...] In den gleichen Zusammenhang fällt gehört der Name für die letzte Vorform des Jazz: Ragtime [...] Das "Zerfetzen der Zeit" durch die Synkope ist ambivalent. Es ist zugleich Ausdruck der opponierenden Scheinsubjektivität, die gegen das Maß der Zeit aufbegehrt, und der von der objektiven Instanz vorgezeichneten Regression."

or whatever other reasons have been marshalled in his defense. Rather, we need to recognize that they make complete sense within Adorno's logic – and in the logic of those who draw from it, defend it, and apply it. Whoever wishes to understand Adorno's cultural critique may easily turn to jazz as his concrete model.

Unlike contemporary exegetes, who usually do not know how to deal with Adorno's analysis of jazz, his own contemporaries were less tentative. In a letter to Adorno, Max Horkheimer expresses favor for his first essay on jazz, which he thinks a "particularly excellent study" of an "apparently insignificant phenomenon." Horkheimer states that the essay was well-received within the Frankfurt School circle for its "formulations of extraordinary precision and brilliance" (qtd. in Paetzold 79, my translation). In his extensive defense of Adorno's writings on jazz, Paetzold writes: "Adorno's analysis of jazz [...] serves as a model for all criticism of the cultural industry, some of which was only carried out later" (77, my translation).[35] We must take Adorno's racial arguments seriously because both superficially and deep-down they make so much *sense* – Adorno's readership, being active sense-makers, consciously or subconsciously abject Blackness along with him. The affective shortcut of anti-Black abjection makes it work. Subconsciously, he can map his argument on Blackness, where the white subject always finds discursive "servility," on *commodified* Black bodies consequently bearing *feigned* subjectivities at best. The metaphor works because Adorno has Blackness labor for it. In this way, we must recognize that, even if it looks like a simplification, the notion of natural Black servility is the affectively axiomatic fixed point of reference from which his argument draws its coherence. His continual reference to jazz as "[n-word] music" or "[n-word] jazz" is often explained or even excused by white scholars as the "language of his time." Such contentions, however, miss the functional role of such vocabulary in the affective dimension of the larger argument. The n-word has been translated into English as "Negro" and later as "black," neither of which sets free the full abjective force Adorno generates by his repeated use of the n-word. However, his lexicality undermines the assumed intellectualist purity of what has been argued to be an aesthetic, musicological argument.

Adorno's jazz critique is white academic abjectorship par excellence because his fascination with Black cultural production is out in the open, revealing the libidinal energy shaped as the desire to deal repeatedly with jazz as well as the symbolic castration of Blackness in order to reassert oneself. Such sexualized language in scholarly prose speaks to the libidinal project of anti-Black abjec-

---

35 "Adorno's Analyse des Jazz [...] dient als Modell für alle, teilweise erst später ausgeführte Kritik der Kulturindustrie."

tion, in which Adorno participates by way of dehumanization and demasculation. His sexualized language is not accidental but functions as a deep-set, libidinal, abjective defense mechanism in view of hypersexualized, phobic Blackness. Adorno was in dire need of a subject-aeffect, which he (presumably) achieved by anti-Black abjection. Ultimately, his defense of a resistant, autonomous subject expressed through art is an ontological dead end. Everything he states about popular culture rings "true-for" white mass entertainment today. There is as little doubt about the marketability of popular culture as commodity as there is a belief in the possibility or relevance of actual resistance. There is a lot of feigned autonomy and pretend individuality in current popular cultural production, regardless of how it is morally judged. This is significant when we look at the Blackness of all popular culture because it appears to imply that Adorno was *right* in his analysis of jazz in the sense that what he sees in jazz from a white position is what any other white mind can see – including those who imitate, capitalize upon, and consume popular culture's Blackness, even though it may not be acknowledged as such. If all popular art derives from a dehumanizing misinterpretation of Black forms, perhaps grounded in the compositional register of high, white art, then an equally dehumanizing critique of these forms would be *correct* in its attack on a generalized popular culture. However, Adorno's critique, even if validated this way, is inconsequential and has lost its grasp on contemporary life. This may sound provocative to those who despise contemporary popular mass culture, who claim that Adorno is more relevant now than ever. I suggest that we must recognize popular culture as a fact, and as a Black one at that. This fact is central to the system of abjectorship by which white subjects desire and live.

The concession of popular culture as a fact (rather than something adversarial to overcome or negotiate with) is the only way we can make this debate at all productive again, which takes us back to the non-relationship between Blackness and Humanity. The anti-Black system of the modern West allows Black people to access the world only as ownable, tradeable commodities. In his argument, Adorno criticizes the commodity for being a commodity so as not to turn into a commodity himself. He can only do so by vibrant anti-Black abjection, by ensuring that the commodity and commodification are elsewhere, namely in Blackness: "With jazz, a disenfranchised subjectivity plunges from the commodity world into the commodity world; the system does not allow for a way out" ("On Jazz" 54). Ultimately, Adorno believes he can save the bourgeois subject by revamping the logic that brought it about in the first place, positing a fundamental, categorical difference and racial hierarchy between autonomous subjects and those who willfully and naturally let themselves be ruled, between 12-tone music and jazz, between Humans and property, whites and Blacks. By engaging in the register of governmentality (including resistance against it), Adorno stays within the logic of anti-Black abjection as he denounces popular culture *for its Blackness*. I suggest that we white scholars should not dabble in analyses of jazz as Adorno does, but consider what the fact of (Black) popular culture says about how we are currently being human. Speaking from the position of white subjectivity, however, Adorno's recognition of the ways in which Blackness is mobilized to cause affect in the popular is spot on. One of the quotations from his jazz critique that tends to meet with a lot of criticism is his statement that "the skin of the black man functions as much as a coloristic effect as does the silver of the saxophone" ("On Jazz" 53). While the condemnation of this sentence is understandable, its racial calling out is significant, and provides part of an answer to Kennel Jackson's question in *Black Cultural Traffic:* "What [...] is it about black cultural material, performances, and representations that puts them in such demand" (19–20)?[36]

The currency of commodified affect

In "The Question of [N***a] Authenticity," Ronald Judy considers the mainstream critique of rap, and by extension hip-hop, as a social malaise that revolves around notions of misogyny, obscenity, violence, and bling bling culture. He dismisses the moral grounds of this criticism by asserting that it assumes Blackness to be a sphere of subjective (and subsequently moral) capacity in the first place. Because this is not the case, moral law does not provide an applicable register for the judgment of Black cultural production, and there is no foundation for demanding a "morally legitimate form of rap" (216). Judy suggests doing away with the superficial "question of [rap's] historical and ideological significance for African American society" on the grounds that this question implicitly seeks to insert a cultural core into the sphere of Blackness, allowing for an optimistic telos such as "the liberation of humans as subjects of knowledge from the subject of experience, from the commodified [n-word] of slavery" (217). Instead, Judy proposes that we recognize the impossibility of a "Black subject" and think of hard-

---

[36] For Jackson, the notion of Black cultural traffic, though it has many forms, "always presupposes movement of cultural matter [and] involves some system of commerce or exchange." He makes sure it is understood that "when we speak of Black cultural traffic, we are always implying the traveling not of whole cultures but elements – even microelements" (8). Thinking about elements – be they microelements, artefactual, or ephemeral – we get closer to the idea of popular culture, in its historical legacy and libidinal structure, as the very Blackness that provides white subjectivity with reinvigorating force and subject-effects of all intensities qua anti-Black abjection.

core rap "*with* the commodified [n***a]," which is a threat "in exhibiting the groundlessness of the sovereign individual" (225). The presence of the hardcore rapper who traffics in the affect they trigger "indicates the identification of human with thing, that the human can only be among things, cannot be beyond or abstracted from things."

In Judy's argument, the [n***a] is a transumption of the "bad [n-word]" not because it signifies criminality (a line of thought Judy describes as "regressive"), but because it opens the possibility of investigating authenticity as such. The "regressive thought" of modern subjectification and governmentality, of political and moral subjectivity, of experience as the ground of Humanity, "cannot comprehend the hard-core [n***a]" because experience as such is "inevitabl[y] los[t] to commodified affect" (227). Whether popular culture is good or bad is irrelevant. These categories belong to the regressive register of white, modern political morality:

> This is the age of hypercommodification, in which experience has not become commodified, it *is* commodification, and [N***a] designates the scene, par excellence, of commodification, where one is among commodities. [N***a] is a commodity affect. [...] A [n***a] forgets feelings, recognizing, instead, that affects are communicable, particularly hard-core ones of anger, rage, intense pleasure. [T]he hard-core rapper traffics in affect not values. (227–28)

This argument would be hard to swallow for Adorno, especially because much of it would, in fact, be intelligible to him. Neither Adorno nor Judy believes (or is interested) in an expressive cultural core, a reducible meaning of Blackness. Both recognize the commodification involved in popular culture as a commodification of psychological and biochemical sensations. Yet speaking at a time when popular culture's hypercommodification is no longer a threat but a fact, Judy's position exposes as regressive Adorno's defense of the bourgeois subject, of political hope (to use Warren's term), of governmentality and resistant opposition, his belief in the possibility of real authenticity, of possibility itself. Unlike Adorno, though, Judy is disinvested in white subjectivity. He has nothing to gain from the idea of autonomous or authentic art, from the notion of Human resistance on political and moral grounds, and so he can theorize popular culture with less emotional judgment. Judy's hardcore rapper, located in the discursive sphere of Blackness, provides a more substantial and workable concept of authenticity *as a tradeable affect*, which ultimately provides the galvanizing ground for popular cultural production, activity, and trade:

> [N***a] defines *authenticity* as adaptation to the force of commodification [...] Authenticity, then, is produced as the value that everybody wants precisely because of the displacement

of political economy with economy [...] Authenticity is a hype, a hypercommodified affect [...] [N***a] is not an essential identity, strategic or otherwise, but rather indicates the historicity of indeterminate identity. (229)

Judy's concept "poses an existential problem that concerns what it means (or how it is possible) to be human" because it rejects modern subjectivity as we know it (229–30).

4 Who Speaks?

4.1 Circulating aesthetics of vitalism

The culture of spontaneity

The earliest improv troupe, The Compass, started up in 1957 at the close of what Daniel Belgrad terms the post-World War II Culture of Spontaneity. At the time, many artistic genres foregrounded a revitalization of the Human based on the ideals of freedom, equality, and the regeneration of the individual. This period, roughly the 1950s and 1960s, is traditionally named and framed as counter-cultural, which positions the discussion and its object in the political sphere of negotiability. To be able to resist in (or even transcend) the political sphere was, consequently, the prerogative of the modern white Man. As an artistic epoch, it thus speaks directly to Adorno's aim as considered in the previous chapter, even though articulated in other artistic and performative forms. According to Belgrad, the Culture of Spontaneity was not "an organized cultural movement but a loose coherence of individually unique artists, writers, and musicians" (5), a loosely circulating set of ideological and aesthetic components, variously applied across art forms and their genres – from writing to painting, from pottery to dance. Belgrad draws lines between the Black Mountain Glyph Exchange and the Beat Generation, between Charles Olsen and Merce Cunningham, Miles Davis, William de Kooning, and Jackson Pollock. He believes that even though "the spontaneous aesthetic avoided politics in the topical sense [...] it was rooted in philosophical concerns that did have political implications" (2). He presents the Culture of Spontaneity as a reaction to corporate liberal "homogenization that rewarded rule-following and attitude management as good in themselves, requiring these qualities in workers as the first step to promotion," and to the increase in mass media that propagated a liberal lifestyle: "Advertisements, mass-circulation magazines, Hollywood movies, and radio and television programs celebrated American technology and the suburban 'standard of living'" (4). Counter-cultural artists sought to oppose, even transcend, such assumed limitations by attempting to foreground an actual, pulsating, breathing reality in their work. The central methodological tool to allow reality to break into the realm of art was spontaneity, believed to bypass the ideological hinderances of a sociopolitical system:

> [T]he conscious mind was the gatekeeper of social proprieties; social alternatives were therefore available first only at the unconscious level. Spontaneous composition avoided the falsifications introduced by a conscious mind that internalized ideological standards.

> By offering unmediated access to unconscious thought processes, spontaneity provided a vantage point from which to question the culture's authority and created the potential for authentic communications exploring new forms of human relatedness. (29)

From this philosophy follows a political claim allegedly inherent in spontaneous art because an "aesthetics of spontaneity, in which authority derives from the artist's ability to consult his or her own unconscious, democratizes access to cultural authority" (40). This logic rests on various axioms and false assumptions that are not self-evident: a) the unconscious mind can be detached from the conscious mind, b) only the conscious mind "internalize[s] ideological standards," c) communication based on the unconscious is more authentic than conscious communication, d) the unconscious is communicable, and e) the unconscious mind exists in some hidden space and hosts universally desirable "social alternatives," and "new forms of human relatedness." In the last assumption, we can see how theorizing anti-Black abjection necessarily undercuts the whole conceptual framework of "spontaneous oppositionality." Once it is recognized that relationality in the political sphere is based on but not inclusive of Blackness, the then-fashionable romanticization of the unconscious as a new stand-in concept for a divine spirit or the soul loses ground.

However, based on these axiomatic grounds, Belgrad suggests that the Culture of Spontaneity saw the emergence of a novel kind of artist figure, "a new type of intellectual, with a different relation to America" (7), suggesting that "socially useful ideas would no longer be articulated in conventional intellectual forms, but would develop new means that did not privilege the abstract intellect" (6). This rebuke of the Cartesian body-mind-duality is central to Belgrad's argument. He foregrounds the revitalization of the Human as Man, the rediscovery of life in its breathing, pulsating form, its actual materiality as an oppositional concept against the stale rationalism of predominant sciences and arts. This dichotomy allows him to sketch out the post-World War II zeitgeist and its (neo-)vitalist philosophical foundations, from which artists in many different disciplines could draw. It is, of course, the same racialized anti-Black dichotomy through which Shusterman and Adorno articulated their theories: intriguing how arguments praising or rebuking popular culture are analogous to (or rhyme with) the romantic self-descriptions of artists and theorists of the Culture of Spontaneity. Like so many things, this parallelism makes sense only if we take anti-Blackness as a galvanizing and structuring force into account. As evaluated by Belgrad, the Culture of Spontaneity was embedded in "a formidable intellectual heritage, including the works of John Dewey, Alfred North Whitehead, and

C.G. Jung, in addition to existentialism, surrealism, gestalt psychology, and Zen Buddhism" (6).[1] All of these appear to challenge the rational subject in one way or another, arguing that there is *something else*, that there is *more to the Human being* than its rationality, that potential and truth are waiting inside the individual subject, which corporate liberalism sanctioned rather than sought. Taking the culture-specific truths-for as universal Truths, artists of the Culture of Spontaneity hoped to re-inspire Humanism, seeking to "develop an oppositional version of humanism, rooted in an alternative metaphysics embodied in artistic forms" (5). Central to this project was "a belief in the value of the unconscious mind as the locus of possibilities denied legitimacy within the prevailing ideology" (15). In this self-positioning outside or against the mainstream thought of rational/political progress and in their reliance on an invisible, mystical, psycho-spiritual entity or force to propel their oppositionalism, these artists had the self-assurance of being *connected to something beyond* the sociocultural mechanics of everyday life.

Racial axioms of vitalism

Vitalism emerged as a reaction to modern biological science, which conceived of human and animal bodies alike as soulless automata. Even though this scientific view of the body is now changing and vitalist thought has gone out of fashion – at least in the academy – it had prominent proponents well into the twentieth century. Vitalist thought assumes that "living organisms are fundamentally different from non-living entities because they contain some non-physical element or are governed by different principles than are inanimate things" (Bechtel and Richardson). Aristotle is usually credited as the first scientist of vitalism in view of his principle of *entelecheia*, later mobilized by biologists like Hans Driesch, who theorized that the development of the embryo represented this principle in human life, "illustrat[ing] the essential difference between the living and the inanimate" (Jones 72). Vitalism has been interpreted and mobilized as "both biologistic and spiritual, naturalist and theological," and its "call to life was a call to restore imagination and creativity against the threat of mechanistic [...] psychology" (Jones 7). Vitalism sought to distinguish the Human subject from the repetition and regularity of mechanical production in both industrial developments and scientific methodologies. Donna V. Jones notes that within mech-

---

1 Shusterman also builds his defense of popular culture on Dewey's ideas of art as experience and philosophy at large.

anistic discourse, *some bodies were more automatic than others*. She states that "the identification of the 'Negro' with the animal or the mechanical (and Descartes had already identified the last two) continues and racializes the usage of 'the mechanical' to express class contempt for repetitive, knowledge-dispossessed (rather than merely unskilled), and hence easily replaceable labor" (35). Later, central vitalist philosopher Henri Bergson refused to "understand the psyche in terms of a mechanical physis in which identical causes yield identical results as positivistically described in precise quantitative law" (29). His philosophy came as a "relief in the age of the machine" (36) – which is not to say it relieved enslaved Blacks from being considered machines with human bodies. Jones's analysis of the conceptual origins and legacies of vitalism is useful here; her underlying thesis is that "one cannot understand twentieth-century vitalism separate from its implication in racial and anti-Semitic discourses" (5). She suggests that "[v]italism was certainly the rage in the early twentieth century, and Henri Bergson was its contemporary prophet" (20), focusing on Bergson because "his philosophy had central categorial importance to European aesthetics and social thought, including its disturbing racialism" (20). She also recognizes Bergson's status as "perhaps the first celebrity philosopher [whose] concepts were ironically taken up in the new networks of mass culture, reduced, popularized and made consumable to an eager and easily bored middle class" (77).

I suggest that the Culture of Spontaneity's rediscovery of vitalist thought follows the same logic as Bergson's contemporaneous popularity. Jones's critique offers insight into the very specific ways improv perceives itself as the progressive betterment of the individual and Human soul, while reenacting and transuming anti-Blackness conservatively and regressively. Jones argues that "Bergson's mnemic vitalism is the opposite of the metaphysics of change that it is understood to be" (21). Based on her analysis, we can link the ideological practice of the Culture of Spontaneity and improv to Warren's critique of a politics of hope as considered above. There is a conceptual relation between the vitalist call to life and the Culture of Spontaneity as an ideological environment in which improvisational theater flourished – it is no surprise that Schiller's idea of being human in play reverberates so powerfully in improv practice and discourse. Examples of vitalist beliefs and aesthetics abound within the Culture of Spontaneity: C.G. Jung's mysticist axioms mobilize the notion of a collective unconscious, setting free the imagination of total human universality and mystical connectedness as "participation mystique" (Belgrad 57); its more philosophical counterparts include "energy" or "force fields" (120), or other ideas of metaphysical "connection" as in Zen-influenced pottery (166). All of these posit an ideal connection to a higher power or self-abandonment – a *giving in to* or *dissolving with-*

in a higher structure, design, or intelligence believed to *inspire* the individual. Regardless of (the lack of) truth-content of vitalist theory, we can safely assert that the Culture of Spontaneity as presented by Belgrad draws directly from vitalist ideas that re-create the Human subject on spiritual or mystical grounds, assuming what Abbagnano describes as "an obscure force [...] that we cannot clearly define [and which] appears to be a close relative of the soul" (118).

In Jones's argument, vitalism becomes an "expression of mysticism" rather than a philosophy or a science (72). She writes:

> The positivists in the natural sciences had little patience for fanciful postulations of a "vital agent"; it goes without saying that many scientists dismissed as vestigial and religious thinking vitalist assertions that any unseen and insubstantial agent might influence the material world [...] Excluded from traditional sciences, vitalist thought flourished in eclectic turn-of-the-century bohemian circles: the occult and alternative social movements. Madame Blavatsky's Theosophical Society, the Rosicrucians, and Aleister Crowley's Hermetic Order of the Golden Dawn had members who were attracted to the broad tenets of vitalism. (73)

The occultism and mysticism of this vitalist strand can be linked to the Culture of Spontaneity's ideology, both in content and in its heedless application. The occultists did not "engage with the details of Bergson's critique of positivism" but only "selected key concepts," most notably the idea of an *élan vital*, "a life force that permeates all things, attainable only through our higher intuitive faculties" (74). Followers of occultism and spiritualism were hooked, and Bergson's concept became the "philosophical evidence of a universal energy that surges through and connects all things [...] with the cosmos" allowing the Human individual to "reach higher planes of consciousness" (74). Jones claims that it was "through these spiritualist movements and not scientific debate that vitalism gained its widest exposure" (75). She further observes that the "popularization of vitalism through the occult underlines a key element of vitalist discourse: the belief in inner and hidden causal factors [that approaches a] determining essence of men." She specifically mentions Madame Blavatsky's focus on the "tapping into great reserves of 'racial memory'" and her "division of the world into a complex racial hierarchy," which, of course, "mirrored the racial hierarchy of the imperial imaginary."

Jones then considers in more detail how "Bergsonian modernism proper emerges out of occultist interpretations and their focus on hidden substances, race memory, and intuition as a privileged state of consciousness" (76). Bergson's popularity arises in part from his reassertion of the Christian matrix in these occultist modern terms, reinvigorating that very matrix by offering up something else, something new that was really the same. Jones writes:

> Because his thought reintroduced the Pauline distinction between an illusory world of solid bodies (including, of course, the flesh) and the impalpable yet truer spiritual world, Bergson spoke powerfully to the crisis of the Catholic Church in the Age of Positivism. Appealing to those exploring spirituality in the nontraditional, occult movements of his time. (78)

(I would argue that postwar neovitalism, identical in its rhetorics and its ideological frameworks, functions similarly.) Bergson, indeed, had no issue claiming that "the only and complete inspiration for universal openness could be found in Christian mysticism and mythology" (81). However, there is a robust axiomatic relationship with previous, distinctly biological vitalism as developed by Driesch. Jones writes:

> Bergson's own philosophy does not clearly guard against a biologically reductionist reading and in fact encourages it at many points. Having analyzed the living being to a thoroughfare through which the impulsion of life is transmitted, Bergson has the individual carry his entire past, a past that extends back to his earliest ancestors and that is augmented with the passage of time. (104)

In his Nobel Prize-winning *Creative Evolution*, Bergson describes memories as "messengers from the unconscious," ponders how "we feel vaguely that our past remains present to us," and mobilizes all dimensions of the past back to "the original bent of our soul" to arrive at the statement that our past "is made manifest to us in its impulses" (5). This is why, for Jones, Bergson's vitalism is *mnemic*. Not unlike Jung, Bergson draws from the assumption of a collective past that activates both a spiritual and a biological dimension, providing the ground for our creativity (particularly the spontaneous kind) and intuition: "The key point for Bergson is that we are free only when our act springs spontaneously from the intuition of the whole continuity of our personality, including our virtual memories, which may include the race's as well, as it has evolved up to the moment of action" (107).

This position is not self-evident and has since been disputed and derailed. Bergson can be interpreted with different foci, for example, in popular affect studies with emphasis on the preverbal. However, Jones addresses a blind spot in the interpretation of what Bergson calls "duration," implying a potentially voluntary ignorance on the part of his interpreters:

> There may seem to be no room for an organic memory with biological and racial resonance or a collective racial memory of which the individual is simply a conduit, but there are clear indications in Bergson's writing that by duration he meant the whole virtual field not only of a single subject's memory but of the race to which he belonged, which now finds its home not in society but on the inside. (110)

With racial memory as the "*sine qua non* of creative spontaneity [vitalism] revolution[ized] man's conception of the past – the discovery of humanity's deep, ethnological time and thus the vast possible store of virtual memory" (114). Jones thereby creates a secure link to the theoretical framework developed for this project: the continuum of the descriptive statements from theocentrism to racialized modernism. She ends by suggesting that "there is a certain isomorphism between a conception of God as an *élan vital* [...] and race not as a fixed essence but as a force that realized itself through ever more complex and powerful concretions" (118). She clarifies:

> My argument is that once the conception of Spirit or God was so revolutionized and dynamized [...] God was soon replaced by race in this evolutionary schema, which we too often equate with social Darwinism. [...] The consequence of the whole evolutionary process was not to have fortuitously created deeply different races; rather the whole point of the evolutionary process is in the first instance to realize various dynamic racial essences. (119)

Accordingly, "once race is understood as the Bergsonian God of the evolutionary process, vitalism is no longer a form of primitivism; it is rather a form of reactionary – nay racial – modernism" (121). We must keep this in mind whenever we encounter manifestations of vitalism – whether in the Culture of Spontaneity or in contemporary improv. When improvisation as an aesthetic mode is regularly celebrated by practitioners and theorizers as a "vital life-force" (Caines and Heble 2), this is the baggage it carries. Vitalism as we know it is historically grounded in the assumption of a collectively shared life-force that connects specific groups. Applying the principle of *entelecheia* to this idea, we would have to deduce: whenever neo-, or post-vitalist ideas emerge, they will always be *intuitive realizations* of this initially racial seed of an idea. Vitalism will always be driven by a life-force, and the *élan vital* for white modern life (and its vitalist celebration) is Blackness.

Obliteration of Blackness

In their choices of method, style, and subject, counter-cultural artists demonstrated the impossibility of revitalizing themselves – white subjects – without recourse to non-white subject matter and methods, part of which was the appropriation of "mere cultures" (Coleman et al. 180; see *III 4.1 Notes on the Rhetorics of Appropriation*). The Culture of Spontaneity provided the vitalist vocabulary grounded in the primitivist fetish by linking Jung's imagined collective consciousness with Native American art. His concept "provided an influential theory linking the method of spontaneous association to the subject matter of primitive myths and symbols" (Belgrad 44). Psychoanalytic theory provided a *legitimate* pathway into the realm of universal truth, in which Native American art was assumed to be inherently and naturally situated:

> The Indians [sic] have the true painter's approach in their capacity to get hold of appropriate images, and in their understanding of what constitutes painterly subject-matter [...] their vision has the universality of real art. (Pollock qtd. in Belgrad 45)

Belgrad takes the easy way out, dodging the violence inherent in the Culture of Spontaneity in a mere paragraph that defends it against cultural appropriation and instead puts forward the notion of "cross-cultural" dialogue. He argues that appropriation is "a clumsy Marxist metaphor [that] reduces all acts of cross-cultural inquiry to the single dimension of theft or dispossession, denying the variety of motives, opportunities, and effects that characterize different modes of cultural exchange" (45). I will not direct my critique at how Native American painters were exploited in the art market of the time or how political decisions ran counter to the fetishized racism directed at Native Americans. Belgrad writes about this without further analysis of the violence inherent in this appropriation of Native Americans, which always makes the white artist a "mythic hero" (61).[2] But while artists of the Culture of Spontaneity sought an Other to vitalize themselves against a system that they felt atrophied their existence in one way or another, to some degree they also conceived of themselves as ethnic outsiders:

> [S]pontaneity was a means for challenging the cultural hegemony of privileged Anglo-American "insiders," giving voice to artists and writers from ethnic and social backgrounds remote from the traditional channels of cultural authority. (15)

Belgrad recognizes the pattern "of a creative artist seeking the means to cultural authority ('looking for a voice' or 'coming to authorship'), who, because of class or ethnic background, begins this search from the disadvantaged position of a cultural outsider" (40). The Culture of Spontaneity arguably performed an absolute political, philosophical, and aesthetic opposition, which speaks to and for the disadvantaged: Charles Olson was affected by "an immigrant father never quite sure of himself," Allen Ginsberg in "Manhattan's upper West Side [...] felt like a shabby, Russian-Jewish interloper from Brooklyn," and Jack Kerouac,

---

2 Consider the apt expression of one Pollock biographer: "What Jackson Pollock derived from Jungian analysis – in addition to a few specific motifs, as opposed to elaborated myths – was permission to engage in his own myth-making" (as qtd. in Belgrad 66).

"whose parents were French-Canadian, felt equally out of place at Columbia University and at the Horace Mann prep school, where he attended on a football scholarship" (Belgrad 41–42).

In terms of the existentialist reality of the Jim Crow era, Belgrad's position displays crass ignorance of the racial realities of the time just before the Civil Rights Movement. However, it serves his aim of drawing a clear-cut and purely aesthetic image of the Culture of Spontaneity without internal ambivalence or differentiation. This might be strategically understandable but does not do justice to the fact that being a "cultural outsider" is different from being a Black-racialized non-sider. While it is not inadequate to think of those artists mentioned as outsiders *on* the sociopolitical grid in the era of Jim Crow, this grid hovered over the sphere of Blackness and provided no positions for Black-racialized artists – except as obliterable reference points. While the gestures of spontaneous appropriation of Native American art were coded in terms of the primitivist fetish, the obliteration of Black cultural production in the form of jazz happened outside the sphere of appropriation. To indulge in Blackness was no transgression; there was no aesthetic "cross-pollination" because Blackness stood in non-relation to the Humanist subject that was being revamped. Additionally, Blackness did not provide the originary but stable meanings that Jung assumed and the artists in the Culture of Spontaneity believed they could find in the content of Native American art. As Wilderson remarks in *Red, White & Black*, "as a Black I have no access to the Indian's spirit world" (46). Rather than providing *meanings* for the Culture of Spontaneity, then, Blackness was drawn upon in terms of libidinality and style. It is widely known, for example, that writers of the Beat Generation – most famously Kerouac and Ginsberg – modeled their art and lifestyle on what they perceived as jazz. Whether in language (from the epoque-making though nonetheless obliterative term of the "beats" to the whole array of "cats" and "squares"), attitude, lifestyle, or art (modeling literary texts after jazz improvisations or structures), jazz served as the abjected ground and Blackness as the vitalizing force from which beat poets existentially drank. Without jazz, they would not have been able to write themselves or their styles into being in the ways they did.[3]

Belgrad talks about what must be read as anti-Black, abjective obliteration by positing bebop jazz as a musical version of the Culture of Spontaneity alongside painting, dance, and ceramics. Slightly longer than a chapter combining ceramics and dance, the jazz chapter is the second-shortest of ten. He writes: "Bebop shared the disposition of other spontaneous art movements at midcen-

---

[3] Norman Mailer's 1957 essay "The White Negro" documents this.

tury to develop an alternative to corporate-liberal culture rooted in intersubjectivity and body-mind holism," articulated through "the African American musical idiom: polyrhythm, timbre, and a structure of call-and-response" (179). With reference to Amiri Baraka's (then LeRoi Jones's) *Blues People*, Belgrad adds that bebop "represented a healthy separatism and autonomy on the part of the black culture" (180). Belgrad's epoch-making does not address the relevance of the 1960s in their potential to challenge "the overrepresentation of Man as if it were human," because the particular "African American" version of his Culture of Spontaneity is framed only as a "recovery of orality [which] *intersected with* the racial politics of the times" (193, emphasis mine).[4] He even claims that "[b]eat poetry and bebop jazz shared a common cultural project: to oppose the culture of corporate liberalism with a spontaneous prosody embodying the tenets of intersubjectivity and body-mind holism" (197). However, these alleged commonalities "led the beat writers to develop strong connections to bebop jazz" (196), and not necessarily vice versa. What he idealizes as romantic "cross-pollination" or mutually beneficial "influence" is always already structured by an anti-Black framework and amounts to little more than the modern white subject drawing and profiting (libidinally and financially) on Blackness. This is even more striking because in the cultural performances and products of the time, it is all laid out in the open, as in Ginsberg's "Howl:"

> Ginsberg himself modeled the structure of his poem "Howl" on the tenor saxophone playing of Lester Young, asserting: "The ideal [...] was the legend of Lester Young playing through something like sixty-nine to seventy choruses of 'Lady Be Good,' you know, mounting and mounting and building and building more and more intelligence into improvisation as chorus after chorus went on." (Belgrad 196–97)

No more were the Beat Generation the only artists to seize on improvisation "as a potent emblem of freedom" (Banes 156). Theaters also *translated* and *transumed* Blackness-as-jazz-as-improvisation, as when taking "the improvisatory structures of jazz as a basis for dramatic form [was] an epiphany for the Living Theater" (157),[5] or when Joseph Chaikin with the Open Theater "refined a particular

---

4 With Godzich, Wynter considers the 1960s a "first phase [...] put in place (if only for a brief hiatus before being coopted, reterritorialized) by the multiple anticolonial social-protest movements and intellectual challenges" ("Unsettling" 262).

5 Banes quotes Julian Beck: "We, who had sought to develop a style through variations of formal staging, found suddenly in the free movement and the true improvisation of *The Connection* something we had not formerly considered. [...] An atmosphere of freedom in the performance was established and encouraged, and this seemed to promote truthfulness, startling in performance, which we had not so thoroughly produced before" (157).

technique of improvisatory sound and movement that he called 'jamming.'" (Chaikin also worked with Viola Spolin's theater games, which I will consider in more detail below.)

In contrast to the appropriation of Native American art, in which the primitivist artist sought symbolic content and universal truth, Blackness-as-jazz was read in purely aesthetic and libidinal terms: jazz structures, improvisation, aliveness, physicality. Black suffering was thereby consumed and performatively extended to the artists themselves – not unlike the way in which the witnesses, spectators, and consumers at the auction block may have *empathized* with the enslaved Africans. While white artists sought to make art like Native Americans, in drawing on Blackness, they superficially adopted styles, language, and modalities of live improvisation and made them their own. This desire-based abjection certainly provided subject-aeffective jouissance even *without generating meaningful content*, of which there is none to be found in the abjected sphere of Blackness or the "banishment from ontology" (Wilderson, *Red* 18). It was the abjective act itself that felt politically good and aesthetically inspirational. Okiji writes that jazz has persistently been "hailed as the bearer of a democratic spirit that is manifest in its inclusiveness, its musical miscegenation, and its rejection of the composer-performer division of labor," and, in a primitivist fashion, jazz-as-metaphor is "abstracted from a black sociohistorical context to serve the needs of a spiritually bankrupt European bourgeoisie" (14–16). The Beat Generation and other artists in the Culture of Spontaneity were living in Blackness-as-metaphor, obliterating those who inhabited it in actuality. Sally Banes also recognizes that Blackness-as-improvisation did not stand alongside, or in relation to, other culturally coded aesthetic modalities at the time: "Improvisation, in particular, was seized on by white avant-garde artists as a potent emblem of freedom. Other traditions were available in the culture [...] But it was the African American tradition, particularly as manifested in jazz, that the avant-garde prized" (156). Blackness must be acknowledged as an "obscure force" and "a close relative of the soul" for the white modern subject engaging in spontaneous artistic practice (Abbagnano 118).

In fact, Belgrad's theorizing itself repeats the anti-Black obliteration that characterizes the Culture of Spontaneity. In passing, he does grant that the writers of the Beat Generation "seem to have remained willfully innocent of the racial power dynamics structuring their reception of the music." However, this line of thought and its consequences are immediately derailed into the politics of hope when Belgrad suggests that "overall they express an excited recognition of bebop as a cultural tool embodying the principles of intersubjectivity and body-mind holism" (210). What is missed is that this very excitement is part of the anti-Black structure that gives rise to what he terms "racial power dynamics."

I have considered Belgrad here because he is one of the first to write up a comprehensive treatment of this aesthetic era, which maintains the presumption of relationality between Blackness-as-jazz and Pollock's appropriation from Native American symbolism. This relationality serves the construction of a coherent theory that explains and encompasses an aesthetic epoch and acts as an academic, conceptual accomplice in its obliteration of Blackness qua abstractification. Anti-Black abjection serves the white scholar-subject to *create* a coherent unity of thought. Such coherence, however, is fundamentally flawed. Belgrad does not theorize in a void. One of his sources is Charles O. Hartman's *Jazz Text*, in which the author comes up with this simplification: "improvisation → spontaneity → genuineness → authenticity → authority" (4). In the subsequent chapters, I will look into the usage of Blackness/improvisation-as-metaphor in more detail. At this juncture, I wish to point out the overall power of Blackness as presented in the blurb of Charles Hartman's publication:

> American arts since World War II have drawn power and mystery from the ideas of voice and of improvisation. These unite in modern jazz, which is America's special contribution to world culture. But American poetry, too, has been vitally motivated by the example of jazz musicians and their ideas of personal sound and spontaneous composition.

Blackness is connoted but not explicitly named in this passage, and this gesture runs through the publication as a whole. There is no cross-pollination, but a ravenous devouring of and violent indulgence in Blackness, neither of which discursively qualify as transgression. There is no *reciprocal relationship* between Blackness/improvisation-as-jazz and whatever else occurred aesthetically in the 1950s and 1960s. (While it is easy to understand what a jazz text might be, like Ginsberg's "Howl," the notion of "text jazz" needs more thought.) And even if we imagine that there might be such relationality, in "many ways, the concerns of the white avant-garde were simply irrelevant to the concerns of black artists" (Banes 158). "Improvisation" has turned "American," replacing minstrelsy as the US American "signature piece" (Kopano 5), while once again offering nothing to those who created, boosted, and lived it. Improvisation has become both an emancipatory practice and a discourse that provides "further evidence of the Slave's fungibility," to use Wilderson's term (*Red* 19). It is laid out in the open how the "figurative capacities of blackness enable white flights of fancy" (S. Hartman, *Scenes* 22). This helps illuminate the power that abjected Blackness holds for the white, progressive, liberal aesthetic practice that is improvisation, which in actuality performs not a universal freedom but a self-aggrandizing universalization of the white modern subject. The obliterative argumentational move from culturally-coded jazz to a generalized improvisation is complicit in anti-Black discourse, be it in artistic practice or scholarly treatments. Employing abstractification to *talk about improvisation* as a modality without meaning (as if anything could be in-the-world without meaning) means that Blackness is continually, obliteratively abjected while the repeated abjective acts of this obliteration reassert the libidinal subject-making aeffect as white. In a sense, improv is to Blackness as the German *digger* is to the n-word: a consciously or unconsciously imagined blackface, a transparent "cloak of Blackness" (Watkins 87).

Abjective phobia and want in improv discourse

"Slave-market bullshit"

Early improv theater stands in a complex fear-and-desire relationship with Blackness. Its method must be read as the obliterative act outlined above; it is ahistorical and willfully ignorant to consider improvisation without recognizing its aesthetic and discursive anti-Blackness. Built on the open-source vulnerability that Blackness provides, early improv was a freewheeling, invigorating mode that anybody and any artistic genre could draw on without being accused of appropriation. Today, this abstractification has been perfected both in the artists and by theorizers; at the time, white performers openly embraced the anti-Black abjective affect that improvisation offered them, expressed in some sort of oppositionality to corporate liberalism and following the same arguments that structured the Culture of Spontaneity at large. But when it comes to metaphorizing anti-Black enslavement as a semantic space of reference for their own "flights of fantasy" (Hartman, *Scenes* 22) while distinguishing themselves as Human subjects from the debased Blackness, improv's founding fathers are a rhetorical class of their own. For Paul Sills – founder of The Compass, the first institutionalized improv ensemble of the era and co-founder of the Second City – "theater is responsible for the image of the human [and] the concerns of the artists are the concerns of the people." In light of corporate liberalism, Sills considers theater in its contemporaneous standard form as "slave-market bullshit" and professional actors as "vestiges of capitalist theatre" (qtd. in Sweet 18). In contrast, he expresses "a love of the authentic, the nonactor, the noble savage, the amateur" (Shepherd qtd. in Coleman 48). To transcend this "slave-market bullshit," Sills mobilizes Blackness-as-improvisation to *better society by way of reasserting a universally human subject* beyond the constrictions of empirical reality:

> [It's about] the awareness that there is such a thing as the self. That the self exists. [...] The authentication of the spirit – which has something to do with the Church – is vital to the theater and is something that the theater can and must do. (Sills qtd. in Sweet 19)

Similarly, improviser Samuel Adams draws on a generalized, metaphorized enslavement stripped of its historical and racial actuality that must be differentiated from the experience of being exploited or alienated from their work (cf. Wilderson 8). He states:

> I think it's a dark hour and everybody better man the pumps and get in there and get ready because this society has turned out an awful lot of slaves – people who are too afraid to move one way or another – and that could cause a lot of trouble. (qtd. in Sweet 17)

The healing of the decidedly white socio-political body through improvisational art, the hope that this is possible through vitalizing the individual who is otherwise "too afraid to move one way or another," positions improv discourse within the conceptual vitalist framework of the Culture of Spontaneity. The adoption of the metaphor and obliteration of the historical reality of Black suffering locates improv discourse in the trajectory of white people imagining themselves as Slave figures to ponder their universalized but culturally specific Humanity.

Remember how Saidiya Hartman considers the abolitionist Rankin, who sought to make Black suffering legible by way of "facilitating an identification between those free and those enslaved" in order to "make their sufferings our own" (Rankin qtd. in Hartman, *Scenes* 18). The same discursive identification is happening here as well, except that no Black-racialized people need to be present for improv's solipsist endeavor. Neither Rankin nor the improviser wants to be treated *like a Slave*. We must keep in mind Hartman's dictum that "in making the other's suffering one's own, this suffering is occluded by the other's obliteration" (19). Broeck builds on Hartman and relates this to eighteenth-century Germany; when

> intellectuals rallied around the metaphor of slavery to push their own claims to self-possession and extension of civil rights, by necessity, they articulated the despicability of submissiveness as their main target. That despicability becomes latched on to the black, who remains irreversibly fixed to slavishness and thus has stood for what the white civilized human is not. The slave, from whom society must be "freed" because her slavishness pollutes and undermines bourgeois sociability, becomes the focus of rejection; "slavery" in its function to symbolize the oppression of humanity, not the white practices of enslavism in which the enlightened bourgeois actively or passively participated, needs to be transcended. This, to 18th century debates, the black in its figuration as the slave becomes useful to the extent that its horrible but distant fate enabled analogical transfer to local scenarios of emancipation from submission to the powers of lordship and nobility. ("Hegelian Maneuvers" 4)

The abjective and obliterative metaphorization of the historically specific figure of the Black Slave is transreal and continual.

Whiteface I: abstractifying the foundational jazz simile
Despite the enslavist idea that Black-racialized people are believed incapable of theatrical improv, and in view of its chronotopical conditions of discursive and institutional autopoiesis, it comes as no surprise that improv has always and continues to be advertised, argued, and theorized with reference to jazz rhetorics on many levels. Here are some examples of the ways in which Blackness is mobilized to talk about improv via jazz:
- Improviser Alan Arkin's description of early Second City: the audiences "understood [improv] and were very excited. It was like verbal-physical jazz [with] the same kind of audience [that] appreciates good jazz musicians" (qtd. in Sweet 225).
- Theater critique Sid Smith in the *Chicago Tribune:* "[A] Harold is comedy in jazz riff. The payoffs come in fits and starts, if at all, and the audiences, sometimes somberly, sometimes ecstatically, gaze in empathy as these adults manufacture like children at play."
- One of the most influential improv groups in Chicago was named "Jazz Freddy" (Kozlowski 53).
- Close et al. in the improv bible *Truth in Comedy:* "A Harold audience will react as if they've seen a Michael Jordan slam-dunk when they watch players remembering each other's ideas and incorporating them back into their scenes" (29).
- A blog post on *Amy Poehler's Smart Girls* dedicated to the likeness between improv and jazz, even reversing the simile: "Like comedy improv, jazz is equally spontaneous; magic is made by an ensemble of artists collaborating around a loose structure and generalized set of rules" (Woods).
- The mobilization of jazz lingua in the discussion of improv theater, as in Wasson:

> Mike [Nichols] and Elayne [May], they made jazz of [the sketch] "Teenagers," by now a standard in their songbook. The dramatic beats of the scene were the melody they riffed around, always a little differently, every time they played it. "You can't plan jazz and you can't plan improv," Nichols said. "They must express you in the moment. You can have your central beats – those are the big laughs and the story points the scene needs – but your breaks have to be there in you and come out." (51)

This selection from improv discourse is arbitrary, but can be (redundantly) added to at will. We also find such Blackness-as-jazz-metaphor in the German context, such as on the website of veteran improv troupe *Emscherblut*, which advertises by stating that "improv theater is like jazz: lively and full of energy" (my translation). Frankfurt's English-language improv Theatre Language Studio similarly advertises its classes in the vein of Belgrad's universalized and celebrated collapse of the Cartesian dualism: "Improvisation breaks down the barrier between mind and voice. It is the theatrical equivalent to jazz music" ("Why"). In providing these examples, my aim is not to attack anyone who applies the simile. Rather I wish to point out that not only can Blackness be seen here in its fungible availability for wanton application; more, it is a *necessary* reference to talk about improv or improvisation at all, mirroring the fact the white subject relies on Blackness to talk about itself from the beginning of modernity into the present.

In academic engagements with improv (or improvisation), the obliteration of Blackness appears to be imperative. Seeking to talk about improvised theater – or improvisation, creativity, and innovation in more general terms – we academics seem unable *not* to draw on jazz to exemplify, decorate, illustrate, or legitimize what is really *our white solipsist* endeavor. Improvisation studies is a vast and expansive field covering disciplines as disparate as philosophy, health, and pedagogy. And wherever one looks, theorization falls back onto jazz – or its second-degree obliterative term, "musical improvisation." Usually, mentioning jazz aesthetics or drawing on examples from the discursive, musical, and historical world of jazz provides the ground and invigoration of that theorization. Keith Sawyer belongs to the first generation of white theorizers on improvisation. His *Improvised Dialogues* is one of the earliest scholarly engagements and was published in 2003 just before improv became the fast and wide-travelling concept it is now. In "Group Creativity" (the essay), he states:

> The study of musical collaboration can provide insights into the study of all group creativity. To make this case, I gave examples of both music and theater group improvisations and I identified the shared characteristics of both types of group creativity. These characteristics are found in all collaboration: in classroom group discussion, in creative domains including art and science and in creative work teams. (99)

In *Group Creativity* (the book), Sawyer devotes an entire chapter to the likeness between improvised theater and jazz ("Jamming in Jazz and Improv Theater"). I don't want to discredit white academics who deal in improvisation in one way or another – this would be grotesque given my own position. My argument is that *whenever anybody* talks about improvisation *as we know it within white abjectorship,* we are most often talking either overtly or covertly about an ahistorical, depoliticized version of jazz, disconnected from the historical and discursive reality that brought it about. The term and concept of "musical improvisation," and higher degrees of obliterative and abjective abstractification like "group collaboration," are complicit in this obliteration of Blackness qua universalization of the mode. Even writing on improv without any interest in "musical improvisation" at all draws on the Blackness of jazz, most commonly via the decorative function of introductory quotes. For example, Chris Johnston opens his *The Improvisation Game* with a quote by Miles Davis, going on to elaborate at length on the universally Human "self" and its potential.

Whether used as decoration, empirical matter, or legitimization, abjected Blackness predetermines everything we say about improvisation. Improv discourse is no exception but it makes rather a strong case in point. It too has always been articulated through jazz. Although almost exclusively white in its demographics (institutions, performers, audience, and performance geographies), its aesthetic ideals and affective configuration are nonetheless fully grounded in Blackness. The modern matrix ensures that this is no ethical dilemma because Black culture holds no locus for cultural production that could be stolen or appropriated:

> As an accumulated and fungible object, rather than an exploited and alienated subject, the black is openly vulnerable to the whims of the world, and so is his or her cultural "production." What does it mean – what are the stakes – when the world can whimsically transpose one's cultural gestures, the stuff of symbolic intervention, onto another worldly good, a commodity of style? (Wilderson, *Red* 56)

In the mode of improvisation, white agents can delve into the sphere of anti-Black abjection via Blackness-as-metaphor. For the white subject, improv is *like* jazz but *better* than jazz because it is not disturbed by the Black realities that accompany jazz as a historical and cultural phenomenon. Improv feeds on and rejects Blackness at the same time. Like the political sphere of the modern West, improv is entirely inspired by, founded upon, and fueled by the very Blackness that it does not allow into the sphere it has created through, with, and on top of anti-Black abjection.

Abstractified even from jazz, improv still makes use of a specifically US discourse that has already cut paths in the national absorption of jazz. Okiji writes:

> The enduring narrative of individuality in jazz sees the music as the mirror of an idealized American society – one founded on the sovereignty of the individual but respectful of the need for concessions that allow for a pragmatic democracy. [...] For the primitivist, jazz presented a course of spiritualized action, a way that decadent modern Europeans, through immersion in the experience of a jazz performance, could be cleansed. Jazz-as-democracy employs the music as evidence of American moral superiority. (16–17)

Okiji addresses the problems inherent in these discursive constructs by discussing how "black America, while contributing to 'democratic symbolic action,' by way of its expression in jazz, poses a direct challenge to the understanding of the terms *America, freedom,* and *democracy*" (17). In the remainder of the chapter, I will show how improv discourse makes use of these argumentational pathways even without needing to refer to Blackness in any overt way. Sam Wasson's *Improv Nation: How we made a Great American Art* elaborates the implicit assumptions and explicit projections of this discourse, already activated even in its title. In a harrowing line of argument, the following excerpt represents the entire network of discursive atrocity that comes with Blackness-as-metaphor when transposed from the field of jazz to improvisation, mobilizing anti-Black all-Americanness. What he writes in the register of political hope is not only a declarative statement of white ownership of Blackness but also – perhaps unintended – an acknowledgment of the depth and laterality to which the Middle Passage, enslavement, and anti-Black abjection have created the modern USA and its subjects on the affective and legislative self-conception of specifically North American US whiteness. It also offers insight into the way Black cultural production can be mobilized to imagine an exclusively white art form as *indigenous to the US* – at once an outward gesture toward Old World Europe and an inward gesture toward those who provide mode and material for national self-assurance, while themselves being precluded from this nationality:

> "Can it be," wrote Kenneth Tynan, [...] "that the European tradition, which regards improvisation as a means to a perfect, fixed and stylised end, is fundamentally inimical to the American tradition, which regards improvisation as an end in itself – as the key, in fact, to a new kingdom of theatrical entertainment? If so, we had better reconsider, for our way of thinking excludes from the theater the kind of invigoration that jazz brought to music." Unimaginable where speech is not free, improvisation is [...] the prodigal son of the First Amendment. [...] It is the tree and fruit of the American mind.
>
> Americans have always been improvisers. "In the language of the Declaration of Independence, for example, Americans accorded themselves the right to revolution, that is, the right to create new forms," writes Professor Kerry T. Burch. "The US Constitution's amendment process similarly codifies permanent revisability as a defining feature of our democratic-inspired political culture." Americans are a work in progress, an ensemble revolution, making it up as we go along. Changeability – the intended imperfection of our foundational documents – opens the way to a more perfect union. And individuals: "Because of the chemistry and the way people were playing off each other," Miles Davis wrote of jazz collaboration, "everybody started playing above what they knew almost from the beginning. Trane [John Coltrane] would play some weird, great shit, and Cannonball [Adderley] would take it in the other direction, and I would put my sound right down in the middle or float over it, or whatever." Through improvisation, they made each other better. Giving themselves over to syncopation, playing, literally, off-beat – term jazz shares with humor – they discover new beats. Surprise and variation, touchstones of improvisation, are requisites of both, amendments – to the melody, the scene, the "law" – permissible only where speech is free. "I think what we look for," explained jazz pianist Bill Evans,

"is freedom with responsibility." Without that freedom, we would be perennially scripted, locked into quarter-note time, unable to evolve new rhythms out of those conflicts that arise from our melting pot morality or, as Del Close described it, the "democratic mess." We can clean it up with improvisation. (84–85)

There is much to unpack here, but in this context, I focus only on those aspects that bring out my argument. First, we must recognize how improv adopts the function of blackface minstrelsy in the nineteenth century in being mobilized for national self-assertion – most explicitly in opposition to Old Europe by drawing on the anti-Black abjective libidinal economy. However, unlike "America's popular culture signature piece," as Kopano describes black-faced minstrelsy (5), the all-American improviser can "indulge the desire to escape the binds of 'civilized' behavior" (Watkins 100) without putting on burnt cork. While also "modeling himself after [...] a black man of [...] natural freedom," the white improviser attains this lively all-American Blackness by the cover of the improvisational mode. The white improviser does not pretend to be Black anymore; the "cloak of Blackness" (Watkins 87), though *aeffectively Black*, has become *discursively transparent* while still enabling (and allowing) the artists "to cast most of their own inhibitions to the wind" (Watkins 87) – which is precisely what people cherish improv for, what they like, why they are willing to spend a lot of money on it. Trading in improv (offering classes and shows) is thus trading in anti-Black affect. In other words, the white improvising subject feels Black without necessarily being aware of it, drawing on Blackness without knowing it consciously. In the popular cultural anti-Black sphere, improv has obliteratively transumed Blackness to the degree that it has become possible to act Black without looking Black. Improv-as-Blackness is a multidimensional representation of the American cultural entity entirely grounded in anti-Black abjection: "Blackness" turned "jazz" turned "musical improvisation" turned "improvisation" turned "Americanness." While keeping the affective experience of Blackness in action, we do not need to examine this logic or its history; we white people have always been in a position to act Black without being Black, but no longer recognize when we're doing so.

Part of what makes the passage from Wasson so disconcerting is that he seems to be well aware of how he mobilizes Blackness. He alludes to Barack Obama's "A more perfect Union" speech from 2008 and (obligatorily) quotes Miles Davis. He misses the fact that the term "off-beat" is not merely *shared* by humor and music, but emerged in a specific cultural context later framed by white scholarship as the Culture of Spontaneity, which in itself performs anti-Black abjective gestures and motions. Like the term "improv," the notion of "off-beat humor" is *secondary to and derivative of* jazz terminology. For Wasson, America (he does not differentiate between the two continents) is as distinctly Black as jazz was for Adorno. While the latter aggressively seeks to destroy American Blackness, Wasson devours it. Where Adorno defends the European subject in his anti-Black abjective argument against jazz and improvisation qua Blackness-as-metaphor, Wasson defends the American subject in his anti-Black abjective argument in a progressively hopeful mobilization of the same concepts. Both run on anti-Black abjection nonetheless. In referencing the Declaration of Independence (both in content as well as inbuilt modality), Wasson willfully and consequentially leaves out historical Black enslavement. It would be too much to ask this author to read the American ideals of egalitarianism and freedom of speech as racialized concepts. Wasson's argument is so blatantly driven by anti-Black abjective jouissance, by the desperate need to be "a good man or woman of one's kind" (Davis qtd. in Wynter, "Unsettling" 271) – that is, a white subject of the United States – that it can almost be taken at face value. We might indeed, as Wasson suggests, understand a universalized improvisation as the "tree and fruit" of the American mind.

Strange fruit, though.

4.2 Intuition and abjection

The concept of intuition features largely in Bergsonian vitalism and its later transumptions in the Culture of Spontaneity and improvisation. Improv performers would rejoice in recognition at statements like this one:

> Our intelligence [...] can place itself within the mobile reality, and adopt its ceaselessly changing directions; in short, it can grasp it by means of that intellectual sympathy which we call intuition [...] To philosophize, therefore, is to invert the habitual direction of thought. (Bergson, *Metaphysics* 69–71)

Jones describes Bergsonian intuition as "absolute in its promise – the transcendence of the seemingly impermeable split between subject and object" (89). Rather than taking up vitalist intuition in its aesthetic-ideological context as a way out of mechanistic, corporate, consumerist, or other restrictions, I look at it from the viewpoints of neuroscience and embodied cognition. Intuition structures our behavior in many situations – for example, when we cannot seem to make a rational choice because there are too many variables, or because we do not have time to weigh the options carefully.[6] Intuition is the gut feeling

---

[6] See further Lehrer's *How We Decide* (2010).

that helps us to act anyway, and to at least assume that we know "the next best thing to do" (Claxton 65). In current usage, by "intuition" we generally mean the ability to make decisions or acquire knowledge without understanding or knowing how we came up with it. It should thus be of no surprise that the concept derives from and (still functions within) an esoteric or religious framework that allows us to think of ourselves (our bodies) as more than we (they) are. If we can know without understanding *how* we know, this logic suggests, we are fundamentally connected to a hidden power or a divine design through intuition. (The unconscious and its mobilization in the Culture of Spontaneity can be read as a transumption of this hidden power.) Many shades of this mysticized intuition can be observed in improv discourse, and in this section, I look at intuition as a generative concept for improv practice and discourse. I consider excerpts of Viola Spolin's originary and influential writings on theater games, and will subsequently argue with Claxton and Damasio for intuition as a pathway to our biopsychological programs rather than a spiritual or moral core, locating it on the sociogenic brink of being human as mythoi and bios. I will develop a conception of intuition fully grounded in the psychosomatic libidinal economy of the body – and thereby immediately related to anti-Black abjection.

Intuition and the unconscious in traditional improv discourse
During the Great Depression in Chicago, pedagogue Viola Spolin worked with "ghetto children from the West side streets" (Coleman 23). In her everyday practice, she developed numerous theater games, which were then collected and published as *Improvisation for the Theater*. The book has been called "seminal" (Coleman 23), and is often referred to as the "bible of improvisational theater" (Spolin back cover). Unlike the contemporaneous and popular Stanislavskian or Strassbergian methods, which were influential in the psychological realism of the stage, Spolin's notion of improvisation focuses on play. It is "basically nonverbal" and "does not emerge from the logical sequence of language, only from the sequence in which objects, the environment, characters, relationships, points of view, and personalities appear out of spontaneous interplay." Each one of Spolin's 222 listed games "pinpoints another outlet off the mainstream where intuition flows – not through the individual mind or ego but through human connections in physical space" (Coleman 27).

The influence of the "high priestess of improvisation"[7] on US improv cannot be overestimated (Coleman 95). Her son Paul Sills used her games in his work with the early Compass; whenever the group felt uninspired, Spolin's workshops reinvigorated them in their theatrical improvisation. What had been developed as children's games became a constant resource for training actors in improvisation. For them to become *stageworthy* involved training and developing their intuition:

> Intuition is often thought to be an endowment or a mystical force enjoyed by the gifted alone. Yet all of us have known moments when the right answer "just came" or we did "exactly the right thing without thinking." Sometimes at such moments, [...] the "average" has been known to transcend the limitation of the familiar, courageously enter the area of the unknown, and release momentary genius within. (Spolin 3)

Departing from the idea of the romantic genius selected by nature, Spolin suggests that, when trained the right way, everybody has the potential of such ingenuity. Such intuition, Spolin continues, "can only respond in immediacy [to] the moment when we are freed to relate and act, involving ourselves in the moving, changing world around us" (4). Her conceptual framework thus speaks to the subsequent assumption by the Culture of Spontaneity that immediacy and spontaneity provide a portal that takes the individual beyond sociality, culture, language, representation:

> Through spontaneity we are re-formed into ourselves. It creates an explosion that for the moment frees us from handed-down frames of reference, memory choked with old facts and information and undigested theories and techniques of other people's findings. (4)

Unlike most of the art Belgrad framed as the Culture of Spontaneity, however, Spolin does not rely on a kind of mystical entity to which one can connect, except the belief in a humanity that we all share. For her, intuition denotes something *inherently human*, "that area of knowledge which is beyond the restrictions of culture, race, education, psychology, and age" (19). This does not mean, however, that Spolin was not sympathetic to esoteric vitalism. In fact, according to Coleman, Spolin was known to "compare her own territory to that of 'Madame What's Her Name:'"

> Blavatsky, the Russian theosophist and mystic [...] shared Viola's view that human consciousness – the spirit – can transcend rational intelligence into a state of divine telepathy,

---

[7] Notably, the head of the iO, Charna Halpern, claims this title for herself as she felt anointed by Del Close to that very position (*Art by Committee* 108).

synergy, oneness and connection with the cosmos, nature, mankind, God, and all the demons, dybbuks, metaphors, and mythologies that throb through the universal mind. (Coleman 30)

Recall Jones's assessment of Blavatsky's "division of the world into a complex racial hierarchy" as a mirror of a Eurocentric racial hierarchy (75–76). Spolin's awareness of and association with this theosophical fetish resonates with the contemporaneous and popular fascination with Eastern philosophy and Western mysticism. In opposition to the dictatorial rule of the "system," it did not matter much whether the way out was Taoist, mysticist, or "the Buddhist thing" (Spolin qtd. in Coleman 32). As a transumption of such theological ideas of *something beyond*, Spolin's rhetoric of awakening develops a mysticism of spontaneous transcendence, of self-discovery through self-abandon. Improvisatory play is construed as a potential pathway to such transcendence, reaching out into the world beyond the game and helping players discover who they *really* are:

> Growth will occur without difficulty in students because the very games they play will aid them. The objective upon which the player must constantly focus and towards which every action must be directed provokes spontaneity. In this spontaneity, personal freedom is released, and the total person, physically, intellectually, and intuitively, is awakened. This causes enough excitation for the student to transcend himself or herself... (Spolin 6)

What in earlier (occultist, spiritual) variants demanded spiritual leadership here becomes a cause and effect logic, whose lacuna around spiritual existence can simply be ignored by imagining a "total personality [...] to emerge as a working unit" (6). What would earlier have been framed as some connection with a spiritual entity becomes the "discovery of the self" (6) or "deeper self-knowledge" (26). These fantasies persist into the present, as in Johnston, who thinks that "the medium of improvisation, properly handled, has the capacity to show us to ourselves" (6).

From the very beginning of Chicago improv, playing theater games originally developed for children was understood to allow adults to return to a realm with which they felt they were no longer connected. Playing these games made them feel they were undergoing mystical experiences; improvisors "experience the practice as a high, a form of bliss" (Coleman 28). Mike Nichols remembers that "once in a while you would literally be possessed and speak languages you didn't speak [...] I don't mean to sound mystical but such things did happen," and for practitioners like Richard Schaal, theater games are "spiritual," "infinite," "everything" (qtd. in Coleman 28). Spolin's system needs no teachers, no authority. At a time when gurus abounded, Spolin did not claim for herself the mystical authority of leadership that earlier variants of vitalist esotericism had. Her playful, improvisatory approach to activate, train, develop, and free human intuition dispensed with all notions of authority: "All words which shut doors, [...] attack the student-actor's personality, or keep a student *slavishly dependent* on a teacher's judgment are to be avoided" (Spolin 8, emphasis mine).

The individual and collective unconscious in improvisation today

Spontaneity and intuition have always been central to improv discourse. The ground for intuitive decision-making has often been called unconscious, celebrated in improv as the sphere of real, originary, and pure creativity. The authors of *Truth in Comedy* suggest that the "unconscious is a lot smarter than most people think," and that "as the players grow more experienced on stage, they discover they have an inner voice which, when followed leads them to interesting twists in the scene":

> The ego is the part of the mind that hangs on to preconceived notions about scenes, so the best improvisers always strive to overcome their own egos. They've learned to trust their inner voices to their unconscious right choices. (Close et al. 91)

Whether we conceive of the unconscious in a traditional psychoanalytic way or apply it as a shorthand for biochemical happenings in the body, it always provides the ground for intuitive action and behavior. Consciousness is assumed to be an obstacle that must be overcome for the improviser to be free to improvise: "Egos have to be sacrificed for the good of the game" (Close et al. 40). The ideal improviser is thus a medium in the strict sense of the term, a "channeler" who articulates content that otherwise lies beyond them. The notion of "flow," which looms large in improv studies, describes a state in which the improviser is no longer bound to the restrictions of consciousness but draws from the pool of unconscious intuition. The term is mainly connected to the work of Mihály Csíkszentmihályi, who uses it to refer to "the holistic sensation that people feel when they act with total involvement" (150). In this state, "action follows upon action according to an internal logic that seems to need no conscious intervention by the actor" (150–51). Acting in flow is thus based on the notion of intuition, in that intuition tells us exactly what to do without rational interference. It provides natural(ized) reactions to a given setting or configuration. If play takes us beyond consciousness, intuition is how we decide when we're in that space of play, and flow is the feeling we have when we play.

While flow is something an individual is assumed to *have* or to *be in*, the notion of "group flow" has been developed by Keith Sawyer to refer to *collective* experiences: group flow sets in when "a group is performing at its peak"

("Group Creativity" 95). Sawyer differentiates his collective-based concept from Csíkszentmihályi's original notion by suggesting that the latter "intended flow to represent a state of consciousness within the individual performer, whereas group flow is the property of the entire group" (95).[8] He continues: "In group flow, everything seems to come naturally; the performers are in interactional synchrony. In this state, each of the group members can even feel as if they are able to anticipate what their fellow performers will do before they do it" ("Group Creativity" 95). Seham too notes that many "players form deep, unspoken connections with teammates" and achieve "a state of unselfconscious awareness in which every individual action seems to be the right one and the group works with apparently perfect synchronicity" (64). In improv lore and discourse, the term "group mind" is commonly used to express this, as described in *Truth in Comedy*:

> After an improviser learns to trust and follow his own inner voice, he begins to do the same with his fellow players' inner voices. Once he puts his own ego out of the way, he stops judging the voices of others – instead, he considers them brilliant, and eagerly follows them [...] When a team of improvisers pays close attention to each other [...] a group mind forms. The goal of this phenomenon is to connect the information created out of the group ideas – and it's easily capable of brilliance. (Close et al. 92)

The iO claims that it holds the key to achieving "group mind," having developed forms that more or less ensure it. The company's owner Charna Halpern claims in *Art by Committee* that "We have the power to [...] provide the audience with a religious experience" (17). This rhetoric plays a large role in iO's perceived cultishness because it resonates with the ideas of spiritual leadership, as can be drawn from this section in Close et al.:

> The [iO] ImprovOlympic workshops constantly prove that a group mind can achieve powers greater than the individual human mind. Scenes created have turned out to be prophetic, and ESP has actually occurred on stage. Players are able to speak simultaneously, at a normal rate of speed, saying the exact same thing, word for word. Some teams become oracles on stage, answering great questions about the universe, one word at a time, leaving the audiences chilled and astonished. Audiences have witnessed the group mind linking up to a universal intelligence, enabling them to perform fantastic, sometimes unbelievable feats. [It] almost seems like they are tapping into the same universal consciousness. (93)

---

8 Sawyer suggests flow can be *owned*. The ambiguity of the term "property" obviously contradicts the idea that flow is unpredictable, ephemeral, and intangible. Sawyer has no issue with the oxymoron "emergent property." However, only the modern white subject is in the epistemological position to own what it previously defined as unownable.

Not many improvers believe in extrasensory perception nowadays, although the trope is still mobilized: "By committing to focusing outward and following the group, we will look to an audience like we have ESP and create that unique improvisational magic" (Gantz). However, the idea of a group mind remains in full effect. For Mark Fotis, it means that "all players [...] are working toward the same goal by opening their awareness and creating one group mind that encapsulates each individual; it is 'e pluribus unum' exemplified" (9). The bloggers of *People and Chairs: The improv blog with attitude* are also fascinated: "Group mind, in my opinion, is one of the coolest things in improv. When group mind is present, you don't steer scenes: you're compelled to move, together. It's about letting go of consciously thinking and being in a state of flow" (Smallwood and Algie).

The idea of support functions as a technique to achieve the higher aim of tapping into a supra-individual unconsciousness because group mind can only be achieved by accepting and agreeing with the ideas of one's fellow players and scene partners. This imperative – though highly problematic – is necessary to produce non-contradictory worlds on the stage. Non-contradictory creativity and creation based on action in flow (that is, on intuitive action) can be understood as the aim of an ideal collective improv scene that also involves the audience.[9] The implications here are obvious, and Seham asks the right question: "When the group works as one mind, whose mind is it? How does the seeming rightness, inevitability, and spontaneity of improv mask the unmarked power of hegemony" (65)? Whatever *truth-for* a group can create or articulate collectively is based on a shared ground of knowledge and ignorance specific to that group. Even Johnston, who otherwise concerns himself with the potential of "another self to be seen, and improvisation [...] conjuring up the exercise" (6), recognizes that the discursive articulation of a group is not necessarily universal. Yet he cannot do without at least hinting at the idea of collective unconscious:

> Improvisation can go further to bring forward themes and images arising from a group working together collaboratively. Arguably, there is a connection here with the notion of a collective unconscious, the existence of a body of shared psychic material that becomes apparent only in its expression. The notion holds particular appeal when a group of people with similar backgrounds commit themselves to an imaginative exercise. The material

---

9 Amy Seham quotes Lisa Trask, "the first woman team coach" at iO:
> I firmly believe, that the audience can be in that group mind too – they're only seconds behind you. We get people screaming "Yes!" because it rings true with them... because they knew it. They don't know how they knew it – but when we said it, it was so. And that's why it's so gratifying for an audience member to see it. [When they] see a good improv show, the audience walks out just as high as the performer does. (65)

emerging from their endeavors can often be identified as saying something about the predicament or life situation of that group. (11)

If a group imagines or experiences itself in group mind or flow grounded in intuitive (or instinctive) behavior, can we reframe that intuition without resorting to the mystical or taking too humble a position toward the unconscious? In improv, the experience of group mind and flow, whether collective or individual, necessitates the mobilization of intuitive behavior – actions we do not consciously rationalize in advance. I suggest we turn to intuition as a core concept to describe improv praxis and see where it takes us. Let me offer another excerpt about group mind from *Truth in Comedy* (Close et al.) which, to my mind, holds more truth than may have been understood or intended:

> There is an empathy among the individuals involved, almost an instinct. The members exist to serve the needs for the group, much like the Inuit Indians who place themselves in a group trance to attack a polar bear or a whale. (92–93)

Here, flow does not serve the emergence of truth via group mind or collective creativity, but instead addresses a culturally specific group activity aimed at destruction and devouring, drawing on a non-white "mere culture" (Coleman et al. 180) to describe that. Such instinctive empathy is directed against a common enemy to kill and enjoy for sustenance, for survival, or – in the case of improv – for fun. Below, I will consider whether and how a culturally specific white intuition can be conceived, and how far this would take us.[10] How do can we productively frame the diffuse concept of the unconscious?

The cultural specificity of white intuition

I am not the first to address the violent romanticism of group mind. Amy Seham raises some important points for debate in her critique of improv:

---

[10] Where improv discourse still has not entirely rid itself of the mysticist framework, of the belief in something higher, in the academic world, the concept of "emergence" has taken the place of collective unity in the Jungian vein. While there might not be a universally shared pool of truthful content, some unknown source, some never-to-be-understood unconscious to be drawn from, the concept of emergence similarly "frees" the individual of individualistic knowledge. German improv scholar Gunter Lösel relies and elaborates on the idea of emergence throughout his *Das Spiel mit dem Chaos*. He takes the notion from Sawyer's *Improvised Dialogues: Emergence and Creativity in Conversation*.

> Because the spontaneous performer seems not to have time to construct images consciously, the social construction of these images seems invisible. Through improvisation, these representations come together, as if by magic, in narratives that appear natural, inevitable, and true, but they are more likely to be drawn from archetype, stereotype, and myth. (xxi)

It is false to assume that spontaneity leads toward "social alternatives," as Belgrad puts it (29). The resonance of several individuals interacting with one another can only appear magical or be experienced as mystical if a shared sociocultural ground already exists. However, I suggest that we take the unconscious more at face value than Seham does. I posit that the emergence of stereotypes is not based on conscious ignorance but represents only the most superficial result of a fundamental libidinal structure also known as the unconscious, which, as Lacan states, is structured like a language. Consider Lyotard in *Libidinal Economy*:

> Theatricality and representation, far from having to be taken as libidinal givens, *a fortiori* metaphysical, result from a certain labour on the labyrinthine and Moebian band, a labour which prints these particular folds and twists, the effect of which is a box closed upon itself, filtering impulses and allowing only those to appear on the stage which come from what will come to be known as the *exterior*, satisfying the conditions of interiority. (3)

Being human, we create theatricality and linguistic representation as the topography of our unconscious, which we need to understand along with Lyotard's extended metaphor as the concrete sociogeny that defines our culturally specific and biochemical being in the world. Our libidinal structure and topography filter external impulses according to the linguistic rules of what has been represented to our consciousness as affective linkages. The selection is based on criteria of "interior satisfaction." We can take this as an elaborate metaphor of how our body-brain functions and how anti-Black abjection partakes in the workings of Lyotard's Moebian band. Considering the unconscious in this way will help us gain a more productive understanding of it.

Intuition helps us short-circuit the complex rationale our body goes through when deciding on "the next best thing to do" (Claxton 10). Our whole body always already (thinks it) knows the best action and behavior in any given situation. Damasio's concept of covert somatic markers ensures that we do not rationalize or judge the possible outcomes of that action, but instead let the body decide what to do. Because the body does so not virtually but in actuality – organs, neurotransmitters, blood pressure, and so on – it feels *natural*, if any interpretable feeling reaches consciousness at all. Very often it does not. However, as Damasio writes in *Descartes' Error*, this "does not mean that the evaluation that normally leads to a body state has not taken place [...] Quite simply, a signal body state or its surrogate may have been activated but not been made the focus of attention" (185). This is how he defines unconscious bias:

> [Triggering] an activity from neurotransmitter nuclei, which I described as one part of the emotional response, can bias cognitive processes in a covert manner and thus influence the reasoning and decision-making mode. (185)

Damasio, then, speaks of intuition when somatic states "operate covertly, that is, outside consciousness" (191). The whole biochemical process that leads to action or inaction in the individual does take place, but it never reaches consciousness. This can make it appear "mystical" at times:

> The explicit imagery related to a negative outcome would be generated, but instead of producing a perceptible body-state change, it would inhibit the regulatory neural circuits located in the brain core, which mediate appetitive, or approach behaviors. [...] This covert mechanism would be the source of what we call intuition, the mysterious mechanism by which we arrive at the solution of a problem *without* reasoning toward it. (191–92)

This is a much more graspable notion of intuition than the esoteric one considered above. Yet the latter still looms large in improv. We must once and for all get rid of the romanticist view that an individual tapping into his or her unconscious will find something objectively truer, something more universal there than anywhere else. A group of individuals will not find something that they do not already know and share collectively – which, on some level, improvisers have always known and stated. Consider Mike Nichols, the first big improv star whose success came with partner Elayne May: "When you have to make things up on the spur of the moment, you gravitate very quickly to the person who understands you most easily" (qtd. in Wasson 37).

Claxton reminds us that intuition is based on experience and is therefore fallible (278). "Experience" here refers to any acquired knowledge or language system and does *not* mean that, for example, phobic Blackness is based on actual experiences with Black-racialized people. Rather, its ground can lie in an observed role model's abjective reaction to Blackness, or it might be learned from the ways in which Blackness is (not) narrativized in the media or other representational spaces. It does, however, involve the fact that Blackness both structures and is being mapped onto the abjective sensation in the body and the way that this sensation, as subject-aeffect, motivates white behavior to become "a good man or woman of one's kind" (Davis qtd. in Wynter, "Unsettling" 271). The learning process begins in the introduction to a semiotic system, to the language that this specific order provides – a process in full swing when the infant starts to *play*. If that semiotic system is ultimately grounded in, structured by, and continually reenacts the dehumanizing praxis of anti-Black abjection, this will necessarily prefigure intuitive responses. The cultural specificity, the *learnedness* of actual biological affectation, is a crucial point in Wynter's "Toward the Sociogenic Principle." She writes:

> But do we, as humans, experience pleasure and satisfaction only from *biologically* appropriate behaviors? Does the opioid system in our case function only *naturally*? If [...] the answer to both of these is a *yes*, then how do we account for the fact, that, as the description of the early seventeenth century Congolese reveal, what was subjectively experienced as being aesthetically "correct" and appropriate by the Congolese (their qualitative mental states of dynorphin-activated aversion on the one hand, and their beta-endorphin activated "pleasure and satisfaction" states on the other) was entirely the reverse of what is subjectively experienced by western and westernized subjects as being aesthetically correct and appropriate? How can the same objects, that is, the white skin color and Caucasoid physiognomy of the Indo-European human hereditary variation and the black skin color and Negroid physiognomy of the African/Congolese human hereditary variation, give rise, in purely biological terms, to subjective experiences that are the direct opposite of each other? [...] Are we not in both cases dealing here with the processes of functioning of two differently culturally programmed opioid systems, two different *senses of the self* of which they are a function? (51–52)

When encountering elements of Blackness that have been somatically marked as abjective spaces or modes and that therefore constitute white people's *primary* affect, providing a ground and projective space for cathecting fear and desire, the white subject-body short-circuits consciousness to figure out intuitively the next best thing to do: abject it.

This is where somatic marking comes in. The active form of the verb reveals that it is not the external object that makes an imprint on the perceiving subject's body. On the contrary, it is the subject's body that marks what it perceives by attributing to it a specific biochemical makeup. Damasio reminds us that "a feeling about a particular object is based on the subjectivity of the perception, the object, *the perception of the body state it engenders, and the perception of modified style and efficiency of the thought process as all of the above happen.*" (147–48, emphasis mine.) Wynter and Damasio intersect here: there is nothing naturally ontological in the object (or subject) as such, but the subject's biological reaction is still real. Phobic Blackness, then, must be seen as a result of the solipsist yet anti-Black abjection of our unconscious when it intuitively decides for us what to do next. This is based on covert somatic marking without representing the activated body loop or neural circuitry to consciousness:

> The explicit imagery [of the chemical body-map] related to a negative outcome [of a scenario triggered by perception or imagination] would be generated, but instead of producing a perceptible body-state change, it would inhibit the regulatory neural circuits located in

the brain core, which mediate appetitive, or approach, behaviors. With the inhibition of the tendency to act, or actual enhancement of the tendency to withdraw, the chances of a potentially negative decision would be reduced. (Damasio 187)

Covert somatic marking supports our *natural avoidance of certain anticipated body states* (such as the dissolution or threatened destruction of one's subject status in the presence of Blackness) and stimulates acting to attain jouissance qua the assertion of one's subject status. The marking is prestructured and preprogrammed in our bodies and comes out as *intuitive action*. Given that white fear and desire are so bound up with anti-Black abjection, white intuition necessarily is too. Abjective reactions to Blackness, driven by fear or desire, are intuitively *correct* for the white subject – and it is of no concern whatsoever what rationalizing that reaction would lead to, or how contemporary superficial morality would judge it. One could go as far as to say that the more anti-Black abjection a white subject enacts, the more *integrity* it has as a subject of the modern West, the more efficient is its body-brain's computational program. These general ideas about locating the cultural specification of white intuition on the grounds of the sociogenic principle and the praxis of abjection may appear counterintuitive or in conflict with the idealized meaning of the term. As discussed above, I suggest we need to think about intuition not as natural but as naturalized, following Wynter's post-Fanonian mobilization of the sociogenic principle as a culturally specific biochemical or psychophysical setup in the body-brain, which conditions us to know or react intuitively to possible solutions to problems of the body-brain's imbalance in favor of psychological as well as physical homeostasis.

What do we make of this in our consideration of improv? I suggest a heuristic differentiation into three fields: a) the intuitive creation of dehumanizing content, b) the unmaking or propertizing of scene partners, mostly while creating that content, and c) the way in which our white modern unconscious biochemically grounded goings-on in the body (*bios*) are structured by a mythological (*mythoi*) dichotomy that divides the world along racial terms into the high and the low, the rational and the irrational, the civilization and nature.

Whiteface II: the abjective ground of being an improviser

We might frame some intuitive, improvisational decisions as white ignorance, but the many instances of Black-racialized improvisers who are not only reduced to stereotypes but propertized, made immovable, or entirely dehumanized on stage speaks to something more. When Dewayne Perkins is approached at the top of a scene with the statement "Now put on this noose," he is not stereotypically depicted as a representative of another culture. He is treated as an enslaved, propertized, thingified, and killable body. When Kimberly Michelle Vaughn is told she will have her "skin stripped off" by white German scientists or Patrick Rowland is "unmade" on stage (both in personal conversation), it is not about stereotypes or misrepresentation. It is about Black bodies used as manifest extensions of a fictional abjected Blackness that the discursive matrix of the modern West holds for them. White intuitive reactions like that do not speak to a lack of rational knowledge, as elaborated by Mills in "White Ignorance." Rather, they reveal the hardwired shortcut for our white subjects' (shared) neural circuitry.

There is a final, more primary dimension here, which I can only conceive of as speculation. At first glance, it appears so *primal* that we can do without theorizing its cultural specificity. What I suggest feels contrived, I think because I too am located within the "cognitive dilemma" addressed by Wynter when she describes how "in the same way as the bee can never have knowledge of the higher-level system that is its hive, we too can in no way normally gain cognitive access to the higher level of the genre-specific autopoietic living system" ("Catastrophe" 32). As an escape, Wynter posits with Césaire "the study of the Word/mythoi," where *mythoi* become an external ground from which we can see how our biological being exists. What I am pursuing here is, in fact, an attempt to follow the consequences of such a study of the Word, deconstruing the *mythoi* that gives rise to our being human as *bios*, which takes me right back to the modern matrix and the chain of transumptive descriptive statements of what it means to be human in historical and contemporary modernity, most prominently the dichotomous way we think of ourselves in the world. I propose that the need for such coherence, the symbolic ground for the biochemical sensations that pertain to the subject-aeffect, as well as the possibility and capacity of attaining it, is a result of the fact that our modern descriptive statement is structured as a dichotomy. This Manicheanism, even if heuristically separated from its racial symbolism, is not natural. It is a semiologically arbitrary fixation of the Human that, drawing on the Platonic postulate to realms of absolutes, sets up an axiomatic stability where there is none. The white modern self is stabilized based on a split psyche that does not exist outside a racial grammar. Therefore, whatever libidinal need ensues from this Manichean split is always already racialized. If, as I argue, the subject-aeffect functions to bridge this split temporarily, there is no subject-aeffect that is not initiated or triggered by an act of anti-Black abjection. Were it not for the Blackness that white modernity has conceptualized as a "state of nature" (Mills, *Contract* 10), we would not be able to conceive consciously (or to dream unconsciously) of the state of nature to which we aspire when playing or improvising. (Again, this is something beyond the cultural phenomenon, the historical and contemporary actuality of improv, which is related to it, but to which it is not reduced.) Were it not for our racialized and abjective libidinal economy, the racialized ground of our psyche, we would not need to experience this specific sensation in this culturally specific way. The grammar would be different. We white subjects would speak, do, and feel different content in a different language. In other words: if the need to improvise – or better, the joy that improvisation brings – did not come from the same place as the desire to be Black and the urge to destroy Blackness, if we did not live in the anti-Black Manichean dichotomy of what it means to be human, improv would not fill this specific need. It would not have such desires projected upon itself. It simply would not be.[11]

From this vantage point, the practice of improvisation no longer refers only to a Black-coded homogenizing descriptor for cultural activity. White performance of improv is no longer obliterative on the grounds of cultural practice. Acting on Manichean anti-Black intuition through the mode of improvisation becomes the very fabric of white modern subject-making. This experience of self is not contingent or coincidental, not *like* but structurally *identical* to the somatic state of anti-Black abjection. We can look to Damasio to endorse the idea that the physical sensation is, in fact, the same:

> It is plausible that a system geared to produce markers and signposts to guide "personal" or "social" responses would have been co-opted to assist with "other" decision making. The machinery that helps you whom to befriend would also help you design a house […] From an evolutionary perspective, the oldest decision-making device pertains to basic biological regulation; the next, to the personal and social realm; and the most recent, to a collection of abstract-symbolic operations under which we can find artistic and scientific reasoning, utilitarian-engineering reasoning, and the developments of language and mathematics. But although ages of evolution and dedicated neural systems may confer some independence to each of these reasoning/decision-making "modules," I suspect they are all interdependent. When we witness signs of creativity in contemporary humans, we are probably witnessing the integrated operation of sundry combinations of these devices. (190–91)

In this logic, any given somatic state may be multifunctional because the same "machinery" works and functions for different emotions. Engaging with (micro) elements of Blackness causes a somatic state for the white subject-body that is

---

[11] Improviser Joel Boyd's statement that "If this issue was solved or wasn't solved [white people's] lives would be the same" would then be wrong (personal conversation). If this issue was solved, we white people would not exist as we do now. But, speaking with Warren, solutions are antithetical to the politics of hope.

identical to or overlaps with the somatic state of play, creativity, or intuition. Because of this overlap, doing improv as a self-experience equates with the somatic state of anti-Black abjection without the subject knowing it. This may be why intuition is so incredibly quick when encountering Blackness, and why people get so creative.

Acting on intuition is thus not reducible to a cultural project of feeding on and obliterating Blackness-as-improvisation or white people monetizing Black cultural production. Neither is it reducible to white indulgence in Blackness-as-metaphor, that is, to white people imagining themselves as Black. Even though Blackness has been abstractified away from these dimensions in some respects, the jouissance of an improv-aeffect can ultimately be traced back to it. This concerns our libidinal hardwiring to the extent that we can be conscious of it. Consider what Newberg and Waldman write about language and consciousness:

> If we want to understand the power of language and human communication, we have to include what we currently know about the nature of conscious thought. Consciousness [...] begins the moment we come out of the womb. Prior to birth the fetus is almost continuously asleep, with very little neural activity occurring in the areas that produce language. [...] Rapid neural growth begins immediately after we are born, as dense neural connections are made between the neocortex, the thalamus, and other deep structures of the brain. (56–57)

Two things are important here. First, it is safe to speculate that when we are developing our affective grammar for the world as infants, we do so not in terms of language, but in modes, modalities, or phenomena: that extra bit of geniality exuded by a friendly uncle when greeting the cousin's Black boyfriend or our mother's flushed cheeks when she comes back from her gospel choir rehearsal. Affective development by way of marking the world, though also linguistic, is prior to the development of verbal language. Second, we need to understand the development of the actual fabric, the physico-chemical structures of our consciousness in reciprocal relation with our affective learning and developing somatic markers *prior to* language, in images, gestures, and sounds. This is a difficult relationship to grasp. If consciousness creates an internal dichotomy between its rational (conscious) self and an abjected sphere marked by non-consciousness mapped onto Blackness, where exactly does our consciousness locate this non-conscious not-self so as to think about it? Within itself or outside itself? Is there any way consciousness can see itself in relation to a non-conscious space other than via the latter's containment, which destroys it? This is the depth of affective language at which we have arrived. I do not know how to further this thought. What I can say, though, is that the white improviser who acts on intuition, on his imagined natural self, his immediate self, unhindered by consciousness, is *consciously* aspiring to a natural, unintellectual, and unconscious (though stable) form of being discursively enshrined in the Black body as abjected space. The aim of white improvisers, whether explicit or implicit, conscious or unconscious, is to generate a somatic state in which they experience themselves as a coherent, unified whole, a smoothly functioning organism that automatically knows the next best thing to do.

I have considered Kristeva's subject-effect, modified as subject-aeffect, at several points in my argument. Up to this point, the reader may have interpreted it as a sensation that can *also* be achieved via anti-Black abjection of any intensity. As a consequence of the above discussion, I now claim that *every subject-aeffect is anti-Black by discursive, affective, subjective necessity.* Kristeva suggests that "out of such straying on excluded ground that [the one by whom the abject exists] draws his jouissance" (8). In the moment of abjective experience, the subject, attracted by the magnetic pull of the abject, lets himself into a state, a sphere, a "land of oblivion" where "the clean and proper (in the sense of incorporated and incorporable) becomes filthy." After and beyond, the abject, returning to the subject, generates a "flash of lightning" that "discharges like thunder" and unifies opposite worlds: "The time of abjection is double: a time of oblivion and thunder, of veiled infinity, and the moment when revelation bursts forth" (8–9). This "bursting forth" provides a kind of pure jouissance that does not merely *correspond to* but *is* the performance of intuition overcoming the split self of the modern white subject. The notion of "whiteface" is instructive here in order to examine white people's activities as they are grounded on our existential dependency on anti-Blackness. The term also highlights that Blackness as white people know it is a solipsist idea that has nothing to do with Black-racialized people. More, it plays ironically on the idea that we have made any progress; we no longer need to put on blackface and openly act or perform Black to experience that abjective sensation in the body. We are no longer even aware of where our doing comes from, what it originates in, what it *means*. (We must be wary of speculations like this one because they may *feel like an intellectual closure on initial questions.* Closure too is a white fantasy.) If we follow this through, play, as "the flow experience *par excellence*" (Csíkszentmihályi 151), offers a fertile ground to theorize this powerful somatic overlap, that is, the simultaneity and mutual reinforcement of overt and covert anti-Black abjection. I will elaborate on this idea in the following chapter.

5 Abjection in Play

5.1 Performing Humanism

In the previous chapter, I engaged with Viola Spolin's application of the term "intuition." Improv historiographer Janet Coleman considers Spolin's work "mystical and intuitive, attached to an energy that is asexual and childlike" (38). The idea of rediscovering your *inner child* through improv, of returning to one's self *before culture*, is a common concept in improv discourse. Improv is thought to help us discover ourselves as we really are, unhindered by social constraints, and the way to get there is through *play* or *playing:* an activity that infants and children are assumed to perform naturally, but one that for adults falls outside the real life of material demands. Playing requires intuitive actions and spontaneous decision-making. Among artists and audience alike, both are believed to lead towards higher authenticity, challenging the social norms that otherwise structure real life and social personas. In the final section of the previous chapter, I suggested that we define intuition as the reactive mode of play in the anti-Black libidinality that makes us, as *adults,* want to play in the first place – and how this subject-making desire is always already anti-Black. In this chapter, I investigate the overlap or identity of the somatic states that make us feel like a coherent subject both in the general arena of play and when performing purposeful acts of anti-Black abjection, focusing on and how these mutually reinforce each other on the improv stage.

Play and playing feature significantly in anthropology and cultural studies, almost always in an idealized performance of Humanism. Schiller's everlasting statement in *On the Aesthetic Education of Man* – "Man plays only when he is in the full sense of the word a man, and he is only wholly Man when he is playing" (80) – is the first theory of play in the Western Humanist tradition. Later, anthropologist Victor Turner theorizes an allegedly universal and all-encompassing *homo ludens*-approach to playing and considers how subjects create culture when at play. Toward the end of the twentieth century, French sociologist Roger Caillois expands the concept of play to encompass social structures and many forms of behavior. He creates a new set of criteria for types and forms of play, such as the differentiation between "a primary power of improvisation and joy," which he terms "paidia," and a "taste for gratuitous difficulty" he terms "ludus." In many ways, improv mobilizes the concept of paidia, so named by Caillois because "it is the root of the word for child" (27). More, improvisers are called "players" more commonly than they are called "actors." The term "game" is used in the discourse in two different ways: Keith Johnstone and

Viola Spolin each developed "games" that prescribe rules for a scene. In certain US-based improv styles, the "game of the scene" is a dramaturgical term to describe the repetition of a pattern, for example a character's behavior. The collocation "free play" also features greatly, most commonly in titles for workshops or publications, such as *Free Play: Power of Improvisation in Life and the Arts* (Nachmanowitch). In addition, the play metaphor in improv connotes the egalitarian ideal of equality. By its nature, improv-as-play is believed to be inclusive and open for everyone:

> Everyone can act. Everyone can improvise. Anyone who wishes to can play in the theater and learn to become "stageworthy." (Spolin 1)

> Improvisational theater is the closest thing you'll find to democracy in the theater. It opens up the possibility of play between the people in the group, and play is an expression of our equality. The crowd I work with regard themselves as equals. (Paul Sills qtd. in Sweet 20)

Improvised play is assumed to provide a space in which everyone can be their truer, deeper, coherent, lively, and more creative self; more, in improv, we can be just like everyone else. The rules and opportunities are the same for everyone, because (according to this logic) in improv we perform our Humanity, and we're all Human. Or are we?

As much as the discourse seeks to present itself as beyond ideological constraints, its liberal racialized idealism plays out the same. In its romantic egalitarianism, improv discourse consistently ignores the fact that different social positions or non-positions rule out the possibility of fairness:

> And why does it have to be that way where it is so segregated, and you need a white man, or you need a white woman, but you don't need more than just one black person's voice. Each of us have different voices. Each minority does. You can't have one minority represent all; that's *unfair* when you have three or four white guys up there. (Vaughn, personal conversation, emphasis mine)

> People do think that everybody is equal. And in some ways they are right. We should be moving towards this era where it shouldn't matter what you look like or what you sound like or where you're from. It shouldn't matter. But it still does. (Joel Boyd, personal conversation)

In an art form based on a romantically universalized but clearly white group mind, the idea of egalitarian play should not be thrown around so carelessly. While improv discourse and theory link theatrical improvisation with play, most of this theorizing a) accompanies self-description and idealized fictions of improv as a cultural phenomenon and, b) seeks to legitimize improv as worthy of academic consideration. My aim is not to reenact the circular or tautological arguments that human play defines humans because humans play. In fact, my discussion can be read against this assumption in that the desire to play as adults is based on the racialized dichotomy through which the modern subject describes itself qua anti-Black abjection. Using a psychoanalytic approach, I relate play in improv with embodied anti-Black abjection and self-making.

5.2 Properties of play

Caillois's concept of paidia "cover[s] the spontaneous manifestations of the play instinct" (28). He continues:

> The elementary need for disturbance and tumult first appears as an impulse to touch, grasp, taste, smell and then drop any accessible object. It readily can become a taste for destruction and breaking things. It explains the pleasure in endlessly cutting up paper with a pair of scissors, pulling cloth into thread, breaking up a gathering [...] etc. Soon comes the desire to mystify or to defy by sticking out the tongue [...] For the child it is a question of expressing himself, of feeling he [sic] is the *cause*. (28)

What enables this "primitive joy in destruction" (28)? And how does it play out in adult play? First, we must recognize that an adult can never go back to their child emotions. The attempt to go back to an imagined child-state, to fancy oneself as transcendent of oneself (while being conscious of it), is a logical impossibility, an uncomfortably loud self-aggrandizement symptomatic of the power of lack in the descriptive statement of the modern Manichean subject.

Donald Winnicott has spent much of his academic and practical career working on children's play, its meanings, and its functions. In my reading, his models and theories link up closely with abjection as a raced, subject-making aeffect, particularly in connection with my previous discussion of Kristeva's abjection in its racialized contemporary realization. According to Winnicott, playing is essential to subject- or identity-formation, and for creativity and the creation of culture. Accordingly, I argue that play becomes a generative space for the adult subject-aeffect, by definition anti-Black. In Winnicott's theory, the original experience of playing takes place in a potential space between the mother-figure and the child. Resonating with Kristeva, the child forms itself in this space through unity with the mother and its repudiation. In this space, the child's body undergoes a sensation of bliss through unity with everything else *as well as* in having its first "not-me" experiences. The undecided oscillation between these (chemical, biopsychical) sensations makes up the potential space of play. Central to Winnicott's play theory is the concept of "transitional objects," through which the child links the realms of reality and fantasy. These objects function as the first possessions of a child and are "both created and discovered [yielding] freedom and joy to babies and all who were once babies" (Winnicott xi–xii). In this section, I briefly sketch out what this entails and link it to abjective jouissance qua dehumanizing, propertizing, anti-Black abjection. I then show how play – both for children and adults – involves aggression and can be related to fantasies of destruction as much as fantasies of self- and subject-making qua ownership claims. Not unexpectedly, we will see that these two poles of play are inextricably linked.

The area of play

Winnicott uses the terms "transitional object" or "transitional phenomena" to designate "the intermediate area of experience, between the thumb and the teddy bear, between the oral eroticism and the true object-relationship." Among these, he includes "an infant's babbling or the way an older child goes over a repertory of songs and tunes while preparing for sleep," and "the use made of objects that are not part of the infant's body yet are not fully recognized as belonging to an external reality." These he calls "first possessions," and they function as symbols by which the fantasy of maternal unity is recreated (2–3). He writes further:

> It is true that the piece of blanket (or whatever it is) is symbolical of some part-object, such as the breast. Nevertheless, the point of it is not its symbolic value so much as its actuality. Its not being the breast (or the mother), although real, is as important as the fact that it stands for the breast (or the mother). (8)

Winnicott focuses on a space in which symbolism and actuality exist simultaneously; transitional objects allow "room for the process of becoming able to accept difference and similarity" (8). He clarifies the paradox that defines the transitional object, which must be accepted rather than resolved or decided:

> The transitional object is *not an internal object* (which is a mental concept) – it is a possession. Yet it is not (for the infant) an external object either. [The] internal object depends for its qualities on the existence and aliveness and behavior of the external object. [...] The transitional object may therefore stand for the "external" breast, but *indirectly*, through standing for an "internal" breast. The transitional object is never under magical control like the internal object, nor is it outside control as the real mother is. (13)

To justify my consideration of Winnicott here, let me clarify a link I see to Kristeva. She too grounds abjection in the infant's experience of breastfeeding or its substitute, and though this is not Winnicott's focus, breastfeeding prefigures the visceral goings-on that develop later when the infant enters the potential space of play. To understand the physico-psychical grounding of play in Winnicott, we must keep in mind the way in which breastfeeding creates the first experience of physical bliss for the newborn. Ignorant of concepts like "hunger," infants learn generically that there is something that will lead to homeostasis, irrespective of where the tension lies. Winnicott writes:

> [A]t some theoretical point early in the development of every human individual an infant in a certain setting provided by the mother is capable of conceiving of the idea of something that would meet the growing need that arises out of instinctual tension. [...] The infant cannot be said to know at first what is to be created. At this point in time the mother presents herself. [...] The mother's adaptation to the infant's needs, when good enough, gives the infant the *illusion* that there is an external reality that corresponds to the infant's own capacity to create. [...] To the observer, the child perceives what the mother actually presents, but this is not the whole truth. The infant perceives the breast only in so far as a breast could be created just there and then. There is no interchange between the mother and the infant. Psychologically the infant takes from the breast that is part of the infant, and the mother gives milk to an infant that is part of herself. (15–16)

The transitional object, then, stands in for the breast in as much as it recreates the illusion that one can create or find an object that will "meet a growing need that arises out of instinctual tension" (16). Whether this object is *created* or *found* is irrelevant: "The important point is that no decision on this point is expected. The question is not to be formulated" (17). (Kristeva's concept of abjection specifies an adult subject's need for occasional, undecided transitionality. Broeck then provides us with a culturally specific cure for this blissful crisis.) To sum up:

> The object is a symbol of the union of the baby and the mother (or part of the mother). This symbol can be located. It is at the place in space and time where and when the mother is in transition from being (in the baby's mind) merged in with the infant and alternatively being experienced as an object to be perceived rather than conceived of. The use of an object symbolizes the union of two now separate things, baby and mother, *at the point in time and space of the initiation of their state of separateness*. (Winnicott 130)

With Winnicott, we can perceive this "time and space" as the area and activity of play, in the simultaneity of internal and external experience realized through the transitional object and accompanied by all its fears and desires. Playing is a mode of exerting magical control on objects. It is not an inner, individual psychic self, nor is it entirely "part of the repudiated world, the not-me" (Winnicott 55). While the latter would be "outside magical control," one cannot play by merely thinking or wishing: "Playing is doing" (55). Rather than locating the area of play in the merely psychical or in the merely external, Winnicott conceives of it as a "potential space between the baby and the mother." This potential space is controlled by the mother-figure who provides the baby with an experience of omnipotence (63). In localizing play, Winnicott makes out two sequential relationships: a) "baby and object are merged with one another," and b) "the object is repudiated, re-accepted, and perceived objectively." If the mother-figure can perform this play

> in a "to and fro" between being that which the baby has a capacity to find and (alternatively) being herself waiting to be found [then] the baby begins to enjoy experiences based on a "marriage" of the omnipotence of the intrapsychic processes with the baby's control of the actual. (63)

It is thus the experience of having and not-having while imagining oneself entirely in control when the having happens:

> Into this play area, the child gathers objects or phenomena from external reality to and uses these in the service of some sample derived from an inner or personal reality. Without hallucinating, the child puts out a sample of dream potential and lives with this sample in a chosen setting of fragments from external reality. (69)

The child experiences having control over these transitional objects, which belong to both the internal and the external sphere.

In theater generally, and in improv in particular, performers are modally forced to treat and read their scene partners as fellow improvising Human beings and – simultaneously – as fictional characters to whom their own fictional characters react. Playing, then, is an apt conceptualization and definition for what happens on the improv stage, but for reasons other than usually asserted. It has much more to do with the players imagining they have magical control over the world *in regard to themselves, that is, their own integral imagination of a world that provides complete integrity:* "In playing, the child manipulates external phenomena in the service of the dream and invests chosen external phenomena with dream meaning and feeling" (69). Improvisers are always interpreting what is happening around them to their own view of what has been happening; they are always reading whatever is happening as "me-extensions" (135).[1] It is thus significant if the adult improviser, in bios/mythoi hybridity, has been preprogrammed to intuitively regard a certain racialized kind of *fungible Black*

---

[1] As Winnicott goes on to tell us: "From the beginning, the baby has maximally intense experiences in the potential space between the subjective object and the object objectively perceived, between me-extensions and the not-me. This potential space is at the interplay between there being nothing but me and there being objects and phenomena outside omnipotent control" (135).

body as a me-extension, in addition to this being a central mode of play. Winnicott links this to Erikson's work on identity-formation, and we can immediately see how the notions of play and subject-making as mobilized in improv discourse relate to anti-Black abjection, to establishing the subject's own sense of coherence within the logic put forward by Afro-pessimist thought.

Transitional Blackness

As I hinted above, playing as adults is not innocent. The anti-Black matrix provides the discursive and libidinal ground for the desired subject-aeffect, which is predetermined in the anti-Black discursive DNA. Blackness provides the frame, the mode, and the material for this experience. This has much more to do with devouring, owning, controlling, and possessing than with equality. In their continual, fungible ownability, Blackness and every signifier carrying its burden are ideal transitional objects for white subjects-in-the-making. White people's play is the culturally specific performance of personal freedom based on the destruction of (Black) transitional phenomena. Even though Winnicott permits us to expand his play theory to a cultural level, it may seem that we are entering dubious epistemological terrain.[2] Like most psychoanalytic theorists, Winnicott is bound to his universalizing gesture, suggesting his theory is valid for all humankind. We can only specify his model by positioning the brackets of a critical approach to Humanism in the first place – as Broeck did with Kristeva's notion of abjection. We must recognize that Winnicott is himself subject to the white ignorance that never ceases to generate fantasies of all-encompassing universalisms. Without challenging the mechanisms described in his play theory, we must specify that he lives in the "racial fantasyland" (Mills, *Contract* 18) of white people, which, by definition, makes itself invisible to its inhabitants. He is speaking from within this fantasy land, and whatever his universalist arguments may gesture toward, he can at best speak about this fantasy world and never beyond. His findings are not wrong; they are culturally specific.

While Winnicott mostly speaks of the (biologized) "mother" or refers to the breast as "part of the mother," on one of two occasions he allows for a less-gendered "mother-figure," leaving it unclear who or what can take the place of the biological mother (55). I suggest that this mother-figure be transumed by the matrix – Latin for "womb" – of white Western modernity, as elaborated by Wynter.

---

[2] Winnicott writes: "There is a direct development from transitional phenomena to playing, and from playing to shared playing, and from this to cultural experiences" (69).

For Winnicott, the mother is responsible for the child's self-experience in the area of play, as defined above: "Playing implies trust, and belongs to the potential space between (what was at first) baby and mother-figure, with the baby in a state of near-absolute dependence, and the mother-figure's adaptive function taken for granted by the baby" (69). Whereas a mother can be "not good enough" – she may fail to provide a background against which a child develops trust in itself in the face of disillusionment (recognizing an external reality) as they transition from one developmental stage to another – this is not the case for the modern matrix of the Western subject. The modern matrix can be said to provide the same trusting space for white play and creativity, better even than the original mother could. We can legitimize the transumption of the mother-figure to mother-culture by recognizing with Winnicott that "this language involving 'the breast' is jargon. The whole area of development and management is involved, in which adaptation is related to dependence" (124). To the very degree to which the modern matrix becomes a culturally specific mother/breast-figure, providing the white subject with magical feelings of omnipotence and ownership over the external world, Blackness becomes the transitional signifier for adults not by choice but by logical necessity (involving both cultural objects and phenomena like songs, ways of behaving, gestures, micro-elements... anything that can carry the meaning of Blackness).

Objects, bodies, and phenomena that signify Blackness fulfill for the white subject all the functions of transitional objects. Most prominently, Blackness triggers the white subject's need (and capacity) for ongoing destruction. To address the linkage between play and destruction, we need to follow Winnicott's differentiation between object-relating and object-usage:

> In object-relating the subject allows certain alterations in the self to take place, of a kind that has caused us to invent the term "cathexis." The object has become meaningful. Projection mechanisms and identifications have been operating and the subject is depleted to the extent that something of the subject is to be found in the object. (118)

If an object is used, the user must understand the object *additionally* as part of an external, shared reality; it is not just "a bundle of projections." Accordingly, "relating can be described in terms of the individual subject" (as a solipsist procedure), while "usage cannot be described except in terms of acceptance of the object's independent existence, its property of having been there all the time" (118). Object-relating and object-usage are by no means contradictory. They denote two different stages in emotional development. Of two infants feeding at the breast, one may be feeding on the self "since the breast and the baby have not yet become (for the baby) separate phenomena," while the other may be feeding on an "other-than-me source, or an object that can be given cavalier treatment" (Winnicott 118–19). Object-relating and object-usage are thus the two stages in between which playing takes place, and which are animated simultaneously by play. Playing is the activity of "feeding from the self" *and* "feeding from an other-than-me source." The same can be said about the solipsist encounter or engagement of white subjects with bodies, objects, or phenomena racialized as Black: it is motivated by and feeds back into the self, but it is dependent on the external existence of sign-vehicles or phenomena. We can read continual anti-Black abjection as play when we think along with Winnicott's argument of the destruction of the object. He suggests that the transition "from relating to usage means that the subject destroys the object." In recognizing the object as an "entity in its own right," the subject was *present in object-relating* with its cathectic qualities, in perceiving it as an external phenomenon rather than conceiving it as a subjective object. This originary, affective understanding of the transitional object is lost, and the object as was is destroyed. Accordingly, the idea of "using an object" is only meaningful in the area of simultaneity as considered here (120).[3]

However, the heuristic differentiation Winnicott subsequently presents is both horrifyingly intriguing and easy to misunderstand:

> [T]here is an intermediate position. [After] the "subject relates to object" comes "subject destroys object" (as it becomes external); and then may come "*object survives* destruction by the subject." [...] A new feature thus arrives in the theory of object-relating. The subject says to the object: "I destroyed you," and the object is there to receive the communication. From now on the subject says "Hullo object!" "I destroyed you." "I love you." "You have value for me because of your survival of my destruction of you." "While I am loving you I am all the time destroying you in (unconscious) *fantasy.*" Here fantasy begins for the individual. The subject can now *use* the object that has survived. (121)

This is an adequate description of the dynamic back-and-forth of white solipsist anti-Black abjection. The object here is not an object qua relationality. Quite the contrary, it functions as a transitional object open to and available for the subject's fantasies of omnipotence. Like Blackness, the transitional object is characterized by its destructibility and expendability, and becomes real/external qua its destruction. In its effect, it is fungible, like sign-vehicles of Blackness. Transitional Blackness thus provides the white subject-body with a means to *meet a*

---

[3] In this argument, we can open up the libidinal grounds of what has been termed white fragility. Somewhere, we white people know (or fear) that Black people are human beings *like us* – with autonomous wills, minds of their own. If we accept that, we are thrown into the vertigo of none-subjectivity; anti-Black abjection as play throws us in and gets us out all at once.

need for omnipotent, magical control, for the integrity and invigoration of the self. In infant play, the externalization originates in the feeling of unity with the mother, of utter bliss and substantial integrity. In adult play, this externalization describes the anti-Black abjective procedure, an affective motion sociogenically inscribed into the subject-bodies of the modern West. Several points in Winnicott's description of the use of an object reverberate powerfully with the abjective modes of fear and desire: the invigoration of the white self, the nonchalant productiveness of destruction, and the idea of Blackness-as-transitional object or transitional Blackness. He writes:

> The object is always being destroyed. This destruction becomes the unconscious backcloth for love of a real object; that is, an object outside the area of the subject's omnipotent control. Study of this problem involves a statement of the positive value of destructiveness. This destructiveness, plus the object's survival of the destruction, places the object outside the area of objects set up by the subject's projective mechanisms. In this way a world of shared reality is created which the subject can use and which can feed back other-than-me substance in the subject. (126–27)

When Winnicott mobilizes *destruction*, he thinks of "the subject creating the object in the sense of finding externality itself, and it has to be added that this experience depends on the object's capacity to survive." He goes on to say – and this is central – that "'survive', in this context, means 'not retaliate'" (120). Even though externalized, the object will not leave the baby/subject. While it represents the baby/subject's lack of omnipotence due to its actual, external, and separate existence, in play it always remains open for abuse. It thus performs that lack and the presence of both lack and omnipotence.

As a floating signifier, Blackness will also always survive without retaliation, even if white play actually destroys a Black-racialized body or object. Blackness will always be there to provide fungible matter: an individual object may be destroyed, but this does not damage transitional Blackness as the structurally abjected sphere, because *any* sign, mode, or gesture meaning Blackness may serve the abjective function that I identify as the same aeffect created by play. And while the white subject needs Blackness to become an external, manipulatable object or phenomenon, that object's or phenomenon's very externality is a threat to the white subject. The white subject is thus both threatened by and dependent on Blackness as a transitional object when play turns into culturally specific abjection. Given this reliance on transitional Blackness, Schiller ultimately (and unexpectedly) turns out to be correct when he states that "man only plays when in the full meaning of the word he is a man, and he is only completely a man when he plays" (80). Because he destroys. Because he abjects. Consider also the uncanny summary Winnicott gives of the special qualities in the relationship between the individual and the transitional object-phenomenon, which reverberates loudly with the functions that Blackness serves for the white modern self: "the infant [white subject] assumes rights over the object, and we agree to this assumption. Nevertheless, some abrogation of omnipotence is a feature from the start" (7). Here are the property claims made by a white subject over Black bodies, materials, and cultural productions *from the start*. While we white subjects agree to this structurally, whiteness has an inbuilt fragility: should we recognize Black-racialized bodies as existent within relationality, that would equate with the subject's abrogation per Winnicott. Further, Winnicott states that "the object is affectionately cuddled as well as excitedly loved and mutilated," and "must survive instinctual loving, and also hating and [...] pure aggression" (7). This involves simultaneous love and hate, fear and desire, destruction and dependence, and the lack of capacity to retaliate that we can observe in white mobilizations of Blackness. The transitional object or phenomenon "must never change, unless it is changed by the infant." This lack of autonomous will on the part of an otherwise externalized transitional Black phenomenon enables the subject to experience the subject-aeffect of omnipotent power over the world and its own being. Moreover, the transitional object "comes from without from our point of view, but not so from the point of view of the baby. Neither does it come from within; it is not a hallucination" (7).

> Its fate is to be gradually allowed to be decathected, so that in the course of the years it becomes not so much forgotten as relegated to limbo. By this I mean that in health the transitional object does not "go inside" nor does the feeling about it necessarily undergo repression. It is not forgotten and not mourned. It loses meaning, and this is because the transitional phenomena have become diffused, have become spread out over the whole intermediate territory between "inner psychic reality" and [...] the whole cultural field. (7)

We are looking at an external object or phenomenon as the source for the internal, biopsychical sensation of the subject-aeffect, an external stimulus for a solipsist procedure.

When Winnicott claims that "playing is essentially satisfying," which is "true even when it leads to a high degree of anxiety" (70), I would add that in its specific anti-Black improv variant, adult play is grounded in excitement and the (omnipotent) control of that excitement. There is an imagined insecurity in not knowing what will happen next and additional insecurity within the white subject-body in the potential (real or imagined) encounter with Blackness. However, I want to suggest that this anxiety is enjoyable in either case, much like Kristeva's horror: "[T]he void upon which rests the play and with the signifier and primary processes [and] the arbitrariness of that play are the truest equivalents of fear" (Kristeva 37). By way of analogy, we can read this along with what happens on the improv stage. We might say that at the beginning, improvisers throw themselves into a situation in which Blackness may potentially denote symbolic life, including the autonomous agency that comes with it, and then *playfully* abject themselves back into order. After the experience of abjective play, the white subject "comes out" cleansed and rejuvenated. For Winnicott, this is the defining feature of playing:

> Playing is inherently exciting and precarious. This characteristic derives *not* from instinctual arousal but from the precariousness that belongs to the interplay in the child's mind of that which is subjective (near-hallucination) and that which is objectively perceived (actual, or shared reality). (70)

Keeping in mind the destructive and devouring aspects that define transitional Blackness, this offers a remarkably succinct description of white solipsist encounters with Blackness, whether on or off stage. Improvisers may seek precisely this anxiety of the unknown when pursuing their craft – the secure but playful look into the abyss of one's own imagined subjectivity, while knowing that ultimately the white subject will always be in control, that our modern mother-matrix will always be there. Can we see this solipsistic violence against transitional Blackness actually play out in an improv scene? Can we translate this violently playful simultaneity of externalizing an internal experience into a solipsist moment, into a theatrical situation of staged improv where Black improvisers are dehumanized beyond the rhetorical scope of misrepresentation and stereotypification? What happens if a player transitionally racialized as Black performs retaliation? Approaching these questions, I will now consider the phrase "treating somebody as a prop" and test its literal meaning by discussing theorizations of property and property rights.

5.3 Improvising property

Meaningful matter: semiotizing the racial improv prop

"I will never play with anybody who will use me as a prop to get a joke off," states Aasia Bullock in personal conversation. The improv prop as a concept is borrowed from the world of scripted theater. In improvised theater, which is celebrated for *not* needing props, "making somebody a prop" means exercising control over one's scene partner, using them to serve and express one's own idea. The body-as-prop exists only insofar as it serves the ideas of the propertizer. It is disabled or discouraged from introducing its own ideas to the scene. Rather than defining the scene partner (or entering the scene oneself) as a tree, a telephone, or a lamppost, "making somebody" a prop denotes a reduction of their creative options, rendering them intellectually inflexible and foreclosing the Human potential they might otherwise display. It is thus a dehumanization that attempts to constrain the scene partner to an unchangeable function in service of the propertizer's scene or story. The prop is treated as fixed bodily matter without even the capacity to control that matter's signification beyond its matterness. In improvisation, then, such *propertization* can be read as the attempt to strip the scene partner of the autonomy so essential to democratic free play. Yet while making somebody a prop undermines the egalitarian ideals of improv, it becomes a key characteristic in play, as defined by Winnicott: as an external phenomenon, the scene partner is manipulated in the service of the manipulator's fantasies and dreams. When treated as a prop, the scene partner serves merely as an external stimulus for the realization of the *truly improvising white agent* doing the propertizing. More than just a cliché, the concept of the prop is highly instructive for three central reasons:

1. It is closely intertwined with intuition as the ground for spontaneous decision-making. It demonstrates individual and collective affective structures of improvisers: a Black-racialized body on the improv stage is always already somatically marked by anti-Black abjective sensations and will be reacted to on this libidinal ground.
2. It is not a metaphor. The racialized treatment of scene partners as props is not about representing stereotypes or the fictionalized reenactment of enslavement but describes the *actual performance of the enslavist, dehumanizing, propertizing regime* that gives rise to the contemporary white subject celebrated in improv.
3. It functions on the aesthetic level of theater semiotics: the propertized improviser (an oxymoron) can be moved, shifted, and *used* for the propertizing white subject's own purposes. This happens on the abjective grounds of (necessarily) anti-Black subjectification and facilitates the communication between a white, Human, improvising cognizer and its Human audience, the referent-we of white improv, without the propertized being itself part of this Human sphere of sociability.

Despite these compelling reasons, the prop in theater has received remarkably little attention in the academy. Scholarly writing on props usually consists of instructions for prop makers, and theater semiotics is only tangentially concerned with props. One exception is Andrew Sofer's *The Stage Life of Props*. As a theater practitioner himself, he offers a multidisciplinary approach in exploring what a prop is, how it works, and what it does on the theatrical stage. Most importantly, he considers how an object *becomes* a prop. Following Sofer, the prop exists at the nexus of materiality and sensual perception on the one hand, and its meanings and signifying functions on the other. Thus it is located right in the center of theater semiotics and instigates debate. Sofer quotes Freddie Rokem as a strong critic of the purely linguistic perspective on the prop. Rokem argues that "the linguistic approach is not able to cope with the fact that even if the object becomes a sign, it never loses contact with its materiality as embodied by that particular object which is present on stage" (qtd. in Sofer 14). Further, "the palates of our mind are stimulated primarily by the chair as a material object and not only as some abstract linguistic food for thought" (15). Sofer specifies the standard OED definition of the prop by defining it as "*a discrete, material, inanimate object that is visibly manipulated by an actor in the course of performance.*" Consequently, being a prop is not an inherent quality of an object, but a quality that an object attains by "Human touch." As Sofer goes on: "It follows that a stage object must be 'triggered' by an actor in order to become a prop," otherwise it will remain decor. Further, "*irrespective of its signifying function(s)*, a prop is something an object becomes, rather than something an object is" (11–12). Sofer draws this description from Jirji Veltruský's article "Man and Object in the Theater." The Prague School semiotician suggests "a fluid continuum between subjects and objects on stage" (qtd. in Sofer 9), so that human bodies can indeed be read and treated as props, granted that they lack an action-force on their own, such as those representing death:

> To use Veltruský's own example, a stage dagger might move from being a passive emblem of the wearer's status to participating in the action as an instrument of murder, and thence a final independent association with the concept "murder." Conversely, when the actor's "action force" is reduced to zero, the actor takes on the status of a mere prop (e.g., a spear carrier or corpse). Actor and prop are dynamic sign-vehicles that move up and down the subject-object continuum as they acquire and shed action force in the course of a given performance. (Sofer 9)

For Sofer, the actor and the prop are two extremes on a continuum, which allows the actor to manipulate the prop (Humanoid or other) to their own ends, while the reverse is not possible. Even though for Sofer (and post-Prague school theater semioticians), the idea of an "'action force' remains murky" (9), in the present context it is quite revealing when read along with improv's Humanist ideals. If we conceive of "action force" as an almost dramaturgical notion, then we can translate it directly into Bergson's idea of an élan vital, a life force, which an audience watches improvisers display and showcase on stage. Such life-force, in improv or elsewhere, corresponds to how whiteness "monumentalizes its subjective capacity, its lush cartography, in direct proportion to the wasteland of Black incapacity" (Wilderson, *Red* 45).

It follows that the theatrical prop is not a phenomenological concept but a functional one, which can be easily related to the equally functional concept of Blackness, that is, to the Black-racialized body as a "will-less actant" in Saidiya Hartman's term (62). The Black body, "seen as paradigmatically *a body*" (Mills, *Contract* 51), is located in the base, earthy, non-homogenous nadir of the earth's matter rather than in the lofty realms of spirits and mind. Like the prop, the Black body is characterized by the perceived foregrounding of its materiality. Without the discursive capacity to harbor a mind, spiritual or rational, it lacks the potentiality of a white subject's action-force. The Black-racialized body on the improv stage is not *treated as* a prop; it *is* the paradigmatic prop for US American entertainment. This analogy, identity even, applies equally to the human body-as-corpse, symbolizing social and material death. Remembering that a prop "never loses contact with its materiality," one technique of making somebody a prop is to do just that: foreground that person's actual body by calling out Blackness. Consider this anecdote from Kimberly Michelle Vaughn:

> I did a jam at Second City recently. It was with teachers and new students. We did a scene, and it was in German, we were all doing German accents and whatever. I was the main person, and apparently, they were scientists. Then they said: "Ok, we're going to strip off your skin, and put a white skin on." These are teachers. White teachers, telling me they were gonna put someone's white skin on me. You could already feel the audience gasping. I was like, "You are not going to strip off my skin as a white person." And the white doctor scientist was like, "No, we're going to strip off your skin because we want you to be no longer insecure. We want you to blend in with society." This is going on on-stage. And then I was going on: "No. It must be so hard to be white, with all that privilege." And one of the other scientists went on: "Oh, it's so hard being vite, because ze guilt, ze guilt..." And then we get a blackout and a proper reaction of laughter. At the same time, I was so mad and wanted to cuss out everyone on stage. But I loved how I stood my ground, regardless of whatever was going on. I was just very uncomfortable. I couldn't believe that teachers at this famous place, Second City, still didn't know how to act in a scene like that. It blew me. I don't think they understood what was going on. (personal conversation)

In this account, the intuitive standard of treating a Black-racialized scene partner exclusively in terms of Black flesh speaks to Wynter's elaboration of the anti-Black matrix, in which Blackness takes the place of pure matter, of the non-moving earth as opposed to the higher (scientific!) reasoning available to white people. On the level of non-egalitarian play, this content was performed over the body of the Black-racialized improviser, whose "no" was white-ignored because its consequences would be unfathomable for the white majority on the stage. In Sofer's words, what happened here for the white referent-we was the performance of a prop "gone awry" (24). He calls this "an instance of autonomy, or pseudoautonomy" on the part of the prop, in the sense that it no longer works according to the set realities of the scene, the rules of the game. It was an autonomous act that was immediately derailed and abjected. The Black improvising body saying "no" was abjected, and its discursive impossibility reinstalled – in much the same way that Adorno discusses his "jazz subject," treating it like the (impossible) "rebellious [n-word]" (Judy 225).[4]

The maintenance of a Black scene partner's status as a prop is invaluable for the white player who lives on his characteristic of being a Human prop-mover. The Black improv performer – a discursive paradox – does not hold the capacity to reverse-propertize a white scene partner. In *Scenes of Subjection*, Hartman writes:

> After all, the rights of the self-possessed individual and the set of property relations that define liberty depend upon, if not require, the black as will-less actant and sublime object. If white independence, freedom, and equality were purchased with slave labor, then what possibilities or opportunities exist for the black captive vessel of white ideality? (62)

In relating the improv scene to real-world behavior, Kimberly Michelle Vaughn displays an optimism of the will and a pessimism of the mind when it comes to collectively unlearn the instinct of anti-Black propertization:

> It's hard when someone keeps making you a prop. Even in situations other than improv. Like in your friends group. All white people, but you need that one black person at the party, right? We're always going to be a prop. But we can't make ourselves a prop anymore. And that's why I think when someone calls us out, we definitely have to make them look stupid for making us that. But it's so annoying that we have to do that. (personal conversation)

---

4 Judy writes:
> A bad [n-word] [...] is an oxymoron: rebellious property. In rebellion, the bad [n-word] exhibits an autonomous will, which a [n-word] as commodity-thing is not allowed to exhibit. There is little more dangerous than a willful thing, through the exhibition of autonomous will, the bad [n-word] marks the limits of the law of allowance by transgressing it [...] The bad [n-word] indicates individual sovereignty, which is to say he is self-possessed. What is at stake here is not the obvious problem of the bad [n-word] embodying the Enlightenment subject (i.e., exhibiting the characteristics of the autonomous subject who is the cornerstone of both civil society and the state). The real threat of the bad [n-word] is in exhibiting the groundlessness of the sovereign individual. Being a [n-word] appearing as a human, the bad [n-word] indicates the identification of human with thing, that the human can only be among things, cannot be beyond or abstracted from things. The bad [n-word] is a human-cum-thing. (225)

Necessary things

That people can be treated as props illuminates how the universalist ethics of free expression and egalitarian play are continuously undermined on a comedy stage. In improv, we have individual Humans on the stage, and these Human subjects act freely. Such personhood is historically bound to propertization in the first place, as legal scholar Margaret Jane Radin argues:

> The premise underlying the personhood perspective is that to achieve proper self-development – to be a *person* – an individual needs some control over resources in the external environment. The necessary assurance of control takes the form of property rights. (957)[5]

Being a person in the first place means having the capacity to structure and control an external reality rather than being controlled and structured by it. Since a subject requires property to exert that control, property must be viewed within a context "that focuses on personal embodiment or self-constitution in terms of 'things'" (958). Radin explains:

> If autonomy is understood as abstract rationality and responsibility, it fails to convey this sense of connection with the external world. Neither does liberty, if understood in the bare sense of freedom from interference by others with autonomous choices regarding control of one's external environment. (960)

The notion of (and experience as) a self is thus "intimately bound up with things in the external world" (960–61). This is highly relevant to the practice of improvisation, where the modern self is showcased as the central attraction on the stage and will thus perform in the mode of manipulative control more often than not. It follows that the abjective moves in the examples I have presented are not glitches or extremes but rather necessities for this white autonomy to exist, performed from the standard default of anti-Black abjection. The German term for prop, "Requisite," (like the English adjective "requisite") comes from the Latin

---

[5] I invite the reader to read the "need for control over resources in the external environment" as a prerequisite for personhood and self-development throughout the entire libidinal register laid out and elaborated upon in the previous chapters, and grounded in Freud's assertion that we draw on externality to satisfy internality. I understand this legal dimension is "on top," or at least "along with" these. Actual ownership must then be understood to provide an abjective subject-aeffect. This aspect is, of course, not presented accidentally at this point of the argument, but must be understood in the primary ownership of those enslaved bodies, Others to Human subjectivity that could hold no ownership and were, consequently, bound to be owned themselves.

requisitum, which roughly means "necessary thing." In terms of improv, this is especially telling; the person who is abjected and owned by the owner-subject is also *existentially necessary* for this subject, that is, for the *freely improvising* actor-agent on the stage. Without this capacity to own, the actor would have nothing to play with (in Winnicott's sense). We might go as far as to say that *there is no improvising subject beyond the precarious prop established early in the scene*. If an improviser feels they are in a state of uncultured nature and unbridled opportunity (as is the – alleged – tabula rasa of the beginning improv scene, not unlike Robinson Crusoe stranded on the island), they will also feel the need to organize it. This is their task as a modern white subject: to establish a property regime on the grounds of Black-racialized abjected bodies.

Property rights and their communication

Contemporary property theory does not conceive of property as something inherent in objects independent of the symbolic exchanges within which that property is traded, and which gives rise to it in the first place. Rather, property is understood in its communicational function – very much like Sofer's functional theory of the theatrical prop. In the early twentieth century, and in the wake of the Marxist denaturalization of property, jurist Wesley Newcomb Hohfeld completely dismissed the idea of property-as-thing and, as Carol M. Rose notes in *Property And Persuasion*, "pointed out that larger entitlements could be analyzed as a series of claims and obligations of varying sorts among persons" (1). Before looking at the communicational dynamics of propertization, we must consider who qualifies for these "larger entitlements" based on the racialized origins of the US property regime itself.

In "Whiteness as Property," legal scholar Cheryl Harris explains how whiteness as such can be framed structurally as a fundamental property required for modern subjective capacity, engaging with the "valorization of whiteness as treasured property in a society structured on racial caste" (1713). She analyzes "the evolution of whiteness from color to race to status to property as a progression historically rooted in white supremacy," based on the observation that "whiteness and property share a common premise – a conceptual nucleus – of a right to exclude" (1714). Modern US ownership has thus always been raced:

> [S]lavery as a system of property facilitated the merger of white identity and property. Because the system of slavery was contingent on and conflated with racial identity, it became crucial to be "white," to be identified as white, to have the property of being white. Whiteness was the characteristic, the attribute, the property of free human beings. (1721)

Harris makes it clear that property must be understood first as a right and second as an object or a thing; it is "characterized as metaphysical, not physical" (1725). She argues that the practice of possession, "the act necessary to lay the basis for rights in property [...] was defined to include only the cultural practices of whites" (1721). Whiteness as a property right ensured that those identified as white could never become somebody else's property, while those racialized as Black could. In a system that enshrines white personhood with freedom-as-unownability, ownership becomes a praxis of self-actualization – always with the enslaved absolutized Human Others in unmediated view, denying them this freedom.

The modern practice of possessing and exercising this racialized and exclusively white right is founded on what Radin calls "property's quintessential moment of chutzpah: the act of establishing individual property for one's self simply by taking something out of the great commons of unowned resources" (9). Rose names two central ways to take possession of property or resources: "(1) notice to the world through a clear act and (2) to reward useful labor" (13). Both are based on traditional axioms of property theory: a Lockean believe that property comes about by mingling one's labor with the thing, and what Rose calls "consent theories," which rest on the assumption that "the original owner got title through the consent of the rest of humanity" (11). Giving notice to the world invariably involves some kind of speech act because it happens in the social mode of communication. According to Rose, this "clear-act principle [...] defines acts of possession as some kind of *statement*. As Blackstone said, the acts must be a *declaration* of one's intent to appropriate" (13). In terms of land, say, this means that whoever lays claim to a plot of land first communicates this claim to the rest of the world, which then gives consent to this firstness by allowing the earliest agent to own that particular plot. Thus, irrespective of ontology, property must be framed as relational: not between the owner and the owned, but between the communicating agency (or agencies) that regulates the property regime of who owns what. Property is the material for relationality. This relationality comes about via communicative speech acts of negotiation:

> Possession now begins to look even more like something that requires a kind of communication, and the original claim to property looks like a kind of speech, with the audience composed of all others who might be interested in claiming the object in question. Moreover, some venerable statutory law requires the acquirer to *keep on* speaking, lest she lose title through the odd but fascinating doctrine of adverse possession.[6] (Rose 14)

---

6 Rose continues: "The doctrine of adverse possession thus transfers property from the title owner to another who is essentially a trespasser, if the trespasser's presence is open to everyone

5.3 Improvising property — 205

To return to improv practice, making somebody a prop involves reading an abject body as an "unowned resource," that is, one that is not in possession of itself (otherwise we would not feel entitled to do so in the first place). Performatively, this procedure involves (speech) acts of laying claim, and these acts request and enable cultural consent. In view of Black-racialized improv props, the claimants can safely count on the intelligibility of Blackness as a vast "unowned resource," descriptively designed especially for the procedure of white modern subject-making and sociability. In the scene related by Vaughn, one of the white improvisers did not respect (ascribe) autonomy to her as an improviser but foreclosed her from the Human fold by drawing on her Blackness as an unowned resource. Vaughn was not given autonomy over (the meaning of) her body. Even though she rejected the skin-change offer, the white, acquiring improvisers in the scene "kept on talking" to ensure continual collective white claimantship over the Black-racialized body. We need to recognize how the white improvisers as scientists felt entitled to make racialized use of Vaughn's presence on the stage. Reduced to Blackness, the fungibility of Vaughn's racialized body was called out: the scene could have been the same with any other Black-racialized performer. Vaughn provided matter and meaning to the scene. Even though she did all she could to preserve a (pseudo?)autonomy, she was not heard when she spoke. Her performance of autonomy was unintelligible for her white scene partners and corresponded to the paradox of the "rebellious [n-word]" (Judy 225).

---

and lasts continuously for a given period of time, and do long as the title owner takes no action to get rid of him during that time" (15).

6 Funny Matter

6.1 Humorous Humanity

When understood merely as an aesthetic modality, improvisation does not need to be funny. However, I am concerned with the specific cultural configuration of US improv, which may be framed as an "artform that stands on its own, with its own discipline and aesthetics," on the one hand, but almost always uses improv and comedy as "synonyms" on the other (Close et al. 14). Including humor in the list of investigative frames of this project alongside intuition and play, I take my cue from the highly influential improv manual *Truth in Comedy*, in which authors Del Close, Charna Halpern, and Kim Johnson "politely tip [their] hats in acknowledgment of the more serious uses of improvisation and saunter off in the direction of chuckles, chortles and guffaws" (14–15). There is ample reason for that; when we look at the curriculum vitae of improvisers who succeed in show business, we see that they have almost exclusively *made it* in comedy, whether hosting, writing for or side-kicking in late shows, starring in comedy movies, or performing on SNL. So there is no reason to detach improv from "the funny," which constitutes a central momentum in its discourse and practice. Improv discourse holds various positions on where "the funny" lives and how to get there. My aim is not to categorize or judge it, but to consider how humor is theorized and what that theorization – including the "observations" on which it is based – tells us about the role and function of humor and laughter for the white modern subject in general, and for improv comedy's anti-Blackness in particular.

Not unlike improvisation studies, contemporary humor studies is a rich academic field. And like improvisation studies, it is an interdisciplinary one: anthropology, social sciences, psychology, philosophy, linguistics, cognitive neuroscience, mathematics, and computational linguistics in artificial intelligence are only some of the fields that have discovered humor as an object of interest. The International Society for Humor Studies (ISHS) is comprised of scholars from such diverse fields as "the Arts and Humanities, Biological and Social Sciences, and Education" and "also includes professionals in the fields of counseling, management, nursing, journalism, and theater." The society studies "humor's many facets, including its role in business, entertainment, and health care as well as how humor varies according to culture, age, gender, purpose, and context" ("International Society"). Yet even with humor theory as an umbrella term for these disciplines, their definitions, terminologies, and methodologies are too diverse to constitute a solid ground for interdisciplinary endeavors. Some talk about the *communicative function* of laughter (signaling play or lack

∂ OpenAccess. © 2022 Michel Büch, published by De Gruyter. This work is licensed under the Creative Commons Attribution 4.0 International License. https://doi.org/10.1515/9783110752748-007

of aggression), others about its *social effect* (degrading those who do not represent the default defined by the laughers). Neuroscientists analyze the biochemistry of laughing (in the release of endorphins) while others discuss *how objects become or why they are funny*. In his dismissal of Hobbesian humor theory, Noël Carroll argues that his approach is "framed in terms of laughter [which undoubtedly] enhances the intuitive plausibility [because] laughter often accompanies triumph," which makes him question whether laughter is "the proper object for a theory of humor" (16). While I appreciate Carroll's differentiation of laughter and humor and his emphasis on humor as a primarily social process, it must be recognized that laughter returns into the concept of humor, for example by way of its communicative function. This social dimension presupposes a sociality in which humor can be performed, and whether scholars are analyzing humor or laughter, the question of being human returns here as a central aspect in the discourse. If neuroscience seeks to discover the biochemical processes behind the elicitation of laughter, this interest is cued by earlier theorists who consider humor, laughter, and the comic not only worthy of study, but central to the study of Humanity as such. The reverse is also true: when mainstream scholarship asks what characterizes Humans (usually in distinction from animals), humor often features as a quality that only Humans possess. Most humor theorists state that "even if it is not the case that humor is a uniquely human phenomenon, it seems to be a nearly universal element of human societies," which is why "it has been a perennial topic for speculation, especially on the part of thinkers ambitious enough to attempt to comment on every facet of human life" (Carroll 6). The result is a myriad of variations on these questions: what are humor's social and individual functions? What makes a given object, text, or situation funny? Are humor or laughter anthropological constants? Is there an anthropological formula that explains how humor works? Is humor linked to what we know as evolution?

This project is not designed to redefine humor studies in general, nor do I claim that these questions are inherently wrong. My point is that in their fundamentally and radically Humanist approaches, humor theorists have not and could never be interested in a critique of that very Humanism. Instead, the Human has been romanticized as a stable and universal referent through which the field has attained and maintains its legitimacy. A fundamental critique – that is, the recognition of Humanity's cultural specificity as opposed to its assumed universality – would challenge not only the legitimacy of the object of study, but the axiomatic setup through which that object attains significance in the first place. The agnotological defense mechanisms are all in order; in most writings that consider race as a factor at all, it is treated as an example of either ethnic (for survival) or derogatory (for superiority) humor, based on degrading stereotypes. We must therefore reassess the argumentative tradition of the rhetorical repertoire mobilized in humor studies. I do so by close reading some central tenets of the humor studies canon as well as the terms of their elaboration. The latter have not only been influential in the making of the discipline as a whole; they also teach us how we know, live, and practice humor as subjects of white supremacist modernity.

Classical humor theory is traditionally differentiated into three categories: superiority theories, release/relief theories, and incongruity theories. On closer inspection, these strands are not separate: there is superiority in relieving laughter and incongruity in the laughter of superiority, while release/relief theories work from both of these. I first draw on Freud's release/relief approach as a central reference, and then consider Henri Bergson's superiority theory, which Freud received favorably. A critical close reading of Bergson's overtly racial argument crystallizes many of strands in this project so far. Finally, I consider the development of so-called "incongruity theories" by tracing the history of Bergson's theoretical trajectory. Early incongruity theories provide the ground for the many different and specific approaches that contemporary humor studies build on today. Here too, by way of their historical trajectory, even the most analytic approaches are always bound to anti-Black dehumanization as the solipsist vector that reasserts the white human subject of modernity. I argue that the term "incongruity" already veils the racially abjective origins of Blackness-as-a-joke in the same way that "musical collaboration" obliterates Blackness-as-jazz.

6.2 Humor and the libidinal economy of anti-Black abjection

The joke and libidinal economy

Freud's humor theory is grounded in and articulated through libidinal economy. In discussing jokes or joke-work, Freud also talks about the mechanics of the unconscious, which motivate joking. Thus, with Freud, one cannot conceive of a joke as purely aesthetic. In *The Joke and Its Relation to the Unconscious*, Freud writes:

> [W]hat we described as joke-techniques – and in a certain sense we must continue to give them this name – are rather the sources from which the joke obtains the pleasure; and we do not feel disconcerted that other procedures should draw from the same sources to the same end. (125)

6.2 Humor and the libidinal economy of anti-Black abjection

Discussing humor via Freud always points beyond "the funny" itself, locating it *along with* laughter and humor in a larger theoretical framework.[1] Freud considers joking as a way to circumvent inner or outer obstacles, the latter standing in the way of the fulfillment of the (sex) drive. Navigating around such obstacles provides jouissance qua avoidance or the prevention of what he calls (with Theodor Lipps) the "psychical damming-up" of libidinal energy. Through humor, and especially in the fictionality of the theater stage, such intentions, such *tendencies*, may find satisfaction. The same is true for internal obstacles like norms and values. In both cases, Freud argues, the gain in pleasure corresponds to the amount of "psychical expenditure" saved rather than spent on the creation or maintenance of such obstacles. He discusses saving psychical expenditure through various technical aspects of jokes, which first set up and then relieve psychical damming-up (114). In view of their similar technical *modi operandi*, Freud asserts a likeness between joke-work and dream-work.[2] He assumes that the function of a dream is analogous to the creation of a joke: "A preconscious thought is given over for a moment to unconscious revision, and the result promptly grasped by conscious perception" (161). This relates humor theory to my discussion of intuition and play. Importantly, however, jokes function *like* but are not the same as dreams for Freud: while dreaming is non-social, joking requires sociality, both in its performative theatricality and in the sense that it *engenders sociality* at large. The communicational and performative features of joke-work distinguish it from dream-work, yet both are psychic processes working toward homeostasis in collective and individual libidinal economy: "The dream predominantly serves to spare ourselves unpleasure, the joke to gain pleasure; but in these two aims, all our psychical activities meet" (*Joke* 173).

Given the role of the unconscious and the desires that libidinal economy creates – specifically the sociogenic libidinal economy of Western modernity, invis-

---

[1] Why engage with Freud's allegedly outdated and overused theory of humor, now mostly treated with historic interest? Whether Freud can provide insight into a higher truth beyond the cultural episteme in which he and I are theorizing is not a question I seek to ask or answer. Nonetheless, I believe it is worthwhile engaging with him without falling for his universalist assumptions, because I find value in the larger theoretical framework of libidinal economy.

[2] Freud writes further: "The interesting processes of condensation with substitute-formation which we have recognized to be the core of the joke-technique in verbal jokes pointed us towards the formation of dreams, for the same psychical processes have been discovered in the mechanism at work here. But that is the very same direction to which the techniques of intellectual jokes also point – displacement, faulty thinking, absurdity, indirect representation, representation by the opposite – and all of these without exception recur in the dream-work [...] Such a far-reaching correspondence as the one between the devices of the joke-work and those of the dream-work will hardly be incidental" (75).

ible to its subjects – we white subjects can never be sure what exactly we are laughing about, or why we are laughing so hard at something. Freud writes:

> [A] good joke makes a general impression of pleasing [Wohlgefallen], so to speak, without being capable of distinguishing directly what part of the pleasure [Lust] comes from the form of the joke and what from its admirable thought-content [Gedankeninhalt]. We are constantly deluding ourselves on this division, now overestimating the quality of the witticism on account of our admiration of the thought it contains, now contrariwise, the value of the joke on account of the amusement we have from how it is clad in a joke. *We do not know what is amusing us or what we are laughing at.* (Joke 126)

The jokers, as much as their audience, are only partly aware of (or morally responsible for) the "funny idea" that crossed their mind, because the origin of the joke is not conscious but intuitive. This is especially significant in the sphere of white spontaneous play with Black-racialized scene partners as transitional objects in the scene:

> It is true, we say, one "makes" a joke, but we sense that when we do so, we are behaving differently from when we make a judgment or an objection. A joke has quite outstandingly the character of a "bright idea," occurring to us involuntarily [...] One senses rather something indefinable, which I would best compare with an *absence*, a sudden letting go of intellectual tension, and then all at once, the joke is there. (Joke 162)

When the subject makes a joke, as in intuitive action, "revelation burst[s] forth," as Kristeva describes the moment of abjection (8). Taking our cue from her statement that "[l]aughing is a way of placing or displacing abjection" (8) we can figure that, though not the same, making a joke and acting on abjective intuition give us white subjects similar sensations. And both lead us to believe that whatever content emerges has somehow mystically occurred to us. A quick look at Freud's original German text is insightful, as it mobilizes a transumed variant of the Reason/Flesh dichotomy, while the English translation mobilizes a Light/Dark dichotomy – both solidly grounded in the modern matrix as analyzed by Wynter. Freud uses the German *Einfall* to modify "idea" in the quotation above, which literally translates to "something that fell into" something else, a Human mind specifically. That is, an idea fell from somewhere into the subject without the latter contributing anything. It fell from *above*, presumably from the sphere of Gods and Reason, an ancient cosmic truth. Translating *Einfall* into English as "bright" instead mobilizes tropes of knowledge and light. The assumption is that the unconscious speaks *through the improviser*, providing "bright" ideas in the same way that God or any spirit of perfect and transcendental knowledge uses the Human subject to speak. The idea of the humorous improviser as a medium or channeler crops up again, echoing also with the vitalist notion that an external force brings spiritual and actual life to matter.

Abjective tendencies in joke-work

In terms of content, Freud introduces two extreme poles: the "innocuous" and the "tendentious" joke. The innocuous joke is an "end in itself," a *blague-pour-blague*, while the latter serves a "particular purpose," thus becoming "tendentious" (85). Unlike the innocuous joke, which is "without content," the tendentious one is "profound" (89). Freud theorizes the innocuous joke in line with Kantian aesthetics as "interesselos" (disinterested), imagining a condition in which "we demand nothing of things, nor wish to do anything with them" (90). The innocuous joke is one of fantasized and fetishized High Reason. (This Kantian vein is obviously racialized, as is Kant's theorizing on the modern Human.) The tendentious joke appears more instructive. With regard to its *libidinal effect* on those making or receiving it, Freud presents the tendentious joke as vastly superior to the innocuous one:

> An un-tendentious joke scarcely ever achieves those sudden outbursts of laughter that make tendentious jokes so irresistible. As the technique can be the same in both, we may find the suspicion stirring that a tendentious joke has sources of pleasure at its disposal – by virtue of its tendency – to which innocuous jokes have no access. (91–92)

With Freud, we can argue that the tendentious joke offers a fulfilling humorous experience not despite but because we white modern subjects do not consciously know what exactly we laugh at:

> Tendentious jokes are able to release [entbinden] pleasure even from those sources that are subject to repression [...] Out of all the developmental stages of a joke, the most important characteristic of joke-work – that it sets pleasure free by removing inhibitions – is most clearly shown in the tendentious joke. It reinforces tendencies it serves by bringing them assistance from impulses kept suppressed, or it puts itself generally at the disposal of suppressed tendencies. (129)

The power of what happens in the depths of the body is precisely what makes tendentious jokes (or the tendentious quality of all joke-work) "irresistible"; unlike the innocuous joke, the tendentious joke has the entire body-brain apparatus working for it. The "pleasure aimed for is not only the pleasure generated by the joke; it is incomparably greater" (131). Freud further distinguishes between two kinds of tendentious joke: the "hostile joke (used for aggression, satire, defense)" and the "obscene joke (used to strip someone naked) [Entblößung]" (92). The hostile joke describes an act of aggression relocated to the realm of the imaginative, that is, the vicarious overcoming of an outer obstacle. The hostile tendentious joke invites its partakers – the performer, the audience – to imagine the destruction of whomever the joke is directed against. Freud thus considers it "well suited to attacking the great, the dignified and the mighty – powers protected from direct disparagement by internal inhibitions or external circumstances" (100).

Superficially, this distinction does not appear to communicate at all with the rest of this project. Modern Blackness is not great, dignified, or mighty, but open for gratuitous and unsanctioned violation. Structurally speaking, no inhibitions or circumstances protect Blackness from attack, and so the humor mode is unnecessary because no obstacle needs to be overcome. In this strict sense, the imaginative, fictional, symbolic reenactment of lynching on the improvised scene is not the overcoming of an obstacle, but merely draws on a cultural repertoire that ensures that Blackness will never signify or represent "hindrance." To make sense of Freudian humor theory in view of the *humorous dehumanization of Blackness*, the hostile tendentious joke must be understood as inclusive of *gratuitously violent destruction.* This particular jouissance comes from the omnipotent control over an externality, and relates to my earlier discussion of play as "transitional Blackness." Acts of humorous anti-Black abjection provide jouissance *in themselves* without needing to be jokes in the strict meaning of the word. The somatic experience of such anti-Black abjection is simply the easiest thing to do for whites. It comes naturally to the modern subject and provides a way to release any tension a nervous improviser might feel on stage. In view of anti-Blackness, then, Freud's notion of an external obstacle to be overcome only translates if we read the mere presence of a racialized, thingified body as an obstacle to (but also as available matter for) Human play as such. In keeping with Winnicott's theory, transitional Blackness on the humor stage functions as a killable vehicle that never goes away and can be drawn upon for *continual destruction.* The tendentious jouissance set free by anti-Black abjection is so powerful and joyful for white jokers and their audiences that it can override the comedian's (assumedly primary) task of crafting jokes.

Whereas the hostile joke is driven by phobia, the obscene joke is driven by desire. Freud analyzes bawdry, which he defines in commonplace terms as the "deliberate emphasizing [of] sexual facts and relations by talking about them [...] directed at a particular person by whom the speaker is sexually aroused" (92). Bawdry talk is an attempt to seduce or to shame. Freud elaborates at length in gender essentialist terms, but one brief section is particularly significant because it refers to his understanding that "the content of bawdry includes [...]

what the two [sic] sexes have in common to which the feeling of shame extends, that is excremental subject-matter in all its range" (93). Freud suggests that at the infant stage, "what is sexual and what is excremental are distinguished badly or not at all" (93). Excremental matter, Kristeva's horror, falls right in the field of abjection: the Fallen Flesh, the existentially abjected not-me of the subject. Blackness signifies physical matter as well as white sexual desire. Like the hostile joke, which can be understood to serve aeffects of destruction, as an example of obscene joking, bawdry talk is a means of attaining certain pleasures in a way that reality would not normally permit: "Bawdry is like an act of unclothing the person [...] at whom it is directed. By voicing the obscene words, it forces the person attacked to imagine the particular part of the body or the act involved and shows them that the aggressor himself is imagining it" (93). In the presence of Blackness, however, the hostile and the obscene joke cannot be distinguished from another because fear and desire aeffect the modern white subject simultaneously. The white subject derives libidinal pleasure precisely qua destruction by calling out Blackness. Recall the skin anecdote related by Kimberly Michelle Vaughn: a group of white male improvisers collectively call out Vaughn for her Black-racialized skin, then attempt to force the scene to move towards surgery to remove that skin. Blackness is treated as a fetish object par excellence, and the white male subjects indulge in exposing it, seeking a non-gendered but sexually arousing subject-aeffect for white people. Blackness functions as the intuitive focal point at the top of the scene, as improvisational play, and provides the theme. In other words, *Blackness fuels a scene that takes the surgical removal of that very Blackness as its theme.* These white improvisers are experts in intuitive, playful, humorous, multidimensional anti-Black abjection. Such propertization of Black-racialized scene partners reduces their function in the scene to an enslaved Black-racialized body to be lynched, or to bodily matter whose skin should be peeled off. It creates a libidinal experience that is both sexual and subject-making, reminding the white subject-body of maternal unity and engendering a subject-aeffect. In terms of anti-Black abjection, then, the obscene and the hostile joke are one and the same. The motivating force of anti-Blackness thus largely structures white humor – especially when performed spontaneously and grounded in intuitive decisions.

Blackness-as-obstacle
One central issue in using Freud's theory of humor is his mobilization of concepts like repression and obstacles, as when he argues that the tendentious joke will "get around restrictions and open up sources of pleasure that have become inaccessible" (98). Anti-Black modernity, however, is structured by the vul-

nerability of and gratuitous violence against anybody racialized as Black, as performed through anti-Black abjection. It is always accessible, so joking would seem unnecessary. I have a two-part response. First, speaking with both Freud and Damasio, we can assert that specific sources of pleasure (somatic configurations) do not serve only one purpose or become activated in only one way. A comparable or even identical release of or relief from tension, of *Abfuhr* or *Entbindung*, can be caused by anti-Black abjection *or* by the mechanics of a joke. In this sense, *Blackness aeffectively becomes the joke* even though the joke may not overtly draw on Blackness-as-content. If it does, its tendentious impetus is all the stronger. This is not a simplified conceptual shorthand; understanding *humorous abjection* as jouissance qua emotional release helps us think this through without relying on the notion of obstacles. If there is an emotional release that provides jouissance qua culturally specific anti-Black abjection, then, second, it can be found in Blackness-as-obstacle, the sexual destruction of which provides a very specific relief indeed: that of an oxymoronic "Black Humanity." This would mean that the maintenance of white subjectivity is in itself libidinally costly; white people need to invest in anti-Blackness for their own existentially subjective pleasure. When we experience the threat of losing that investment, we face an obstacle between ourselves and the reward for our investment. Considered from this angle, Blackness is the ultimate obstacle. The culturally specific sociogeny of the modern West relies on the Blackness that simultaneously threatens it, because the recognition of "Black Humanity" would dissolve the descriptive statement of what it means to be Human. As white subjects we relieve ourselves of the tension caused by this oxymoron via anti-Black abjection. This abjective praxis ensures that only white people can be Human, even if the concept's stability requires the ongoing dehumanization and continual destruction of Blackness.

Improv, as an exclusively Human activity, is foreclosed to those racialized as Black who are read as "talking bodies" at best, as Charles Mills phrases it: "Whites may get to be 'talking heads,' but even when Black heads are talking, one is always uncomfortably aware of the bodies to which they are attached" (*Contract* 51). When Black-racialized bodies do what they do not have the discursive capacity to do, even when they simply move in a sphere in which moving is what Humans do, they *represent* an obstacle. Blackness is thus *in the way* of real (undisturbedly abjective white) fun, and *a central constituent* of the humorous situation. It is transitional for the white subject who delves into abjective transitionality in humorous play. Winnicott's description of the subject-in-transition speaking to its object is again relevant: "'You have value for me because of your survival of my destruction of you.' 'While I am loving you I am all the time destroying you in (unconscious) *fantasy*" (121). Watkins writes plainly that "[s]laves as comic figures were [...] both the vehicle for and butt of the humor" (62–63).

Hiding Wynter's referent-we in Freud's "third person"

According to Freud, "a tendentious joke requires three persons: apart from the one who is telling the joke, it needs a second person who is taken as the object of the hostile or sexual aggression, and a third in whom the joke's intention of producing pleasure is fulfilled" (*Joke* 95). A joke only exists in performative actuality, which it shares with the theatrical situation. Both rely on a configurative setup in which performers do something while being watched by an audience. Only in this setting can both fulfil their communicational and performative functions, can they *make sense*. This tripartite configuration also resonates with Rose's elaboration on property claimantship; after the owner's initial "moment of chutzpah: the act of establishing individual property for one's self simply by taking something out of the great commons of unowned resources," they must give "notice to the world through a clear act" (Rose 9, 13). The theatrical situation is defined by the attentive presence of an audience. In view of joking, Freud notes: "the pleasure produced by the joke turns out to be more evident in the third person than in its author" (*Joke* 140). Parallel to the speech act of a property claim, which is successful only when a group (or audience) consents to the claim, in humor the laughers ultimately realize the joke and mark its success or failure. The performance is only completed with the so-called third person's libidinal reaction – regardless of how or whether that reaction is expressed. (I would suggest that it only needs to be imagined by the joker, making the third person a virtual entity.) This is existential for the theatrical situation of joking. Without an (imagined) audience, an (imagined) shared communality, a joke not only *achieves* nothing: it does not exist.

In distinction from the comic, Freud notes the crucial (potentially imaginary) sociality of the tendentious joke:

> If the joke puts itself at the service of hostile tendencies or intentions to strip [expose] someone, it can be described as a psychical process requiring three persons [...] the psychical process of the joke is consummated between the first person, the 'I,' and the third, the person from the outside. (139)

This reads at least as an analogy if not an identical procedure to anti-Black sociogeny. If we read theatrical joke-work not as technique but as affective meaning or knowledge-making that turns mythoi into flesh, it becomes existentially dependent in this very sociality. The function of the joke, then, can only be realized on the grounds of an (imagined or real) community. The performative effect of a joke relies on the existence (or imagination) of a group in which this culturally specific libidinal tension-relief set up can operate intelligibly. The (imagined) group must share the same libidinal ground:

> [The third person] must definitely be compatible psychically with the first person to the extent of sharing the same internal inhibitions that the joke-work overcame in the first. [...] Every joke demands its own audience, and laughing at the same jokes is evidence of far-reaching psychological compatibility. (145)

This is how theatrical, tendentious jokes participate in the hardwiring of the Western subject's body-brain. Freud understands humorous communication as the process of re-assertion qua audience feedback. Working from the observation that one would not laugh at one's own joke, he speculates on why subjects feel the urge to tell the joke to a third person in the first place:

> We can only surmise [...] that the *very* reason we are compelled to pass on our joke to someone else is because we are unable to laugh at it ourselves. From our insights into the conditions for gaining pleasure and for release in the third person, we may infer that in the first the conditions for discharge are lacking and those for gaining pleasure are perhaps only incompletely fulfilled. If so, it is not implausible that we supplement our own pleasure by achieving the laughter that is not possible for ourselves by the roundabout way of the impression on the third person who has been made to laugh. (149)

Telling a joke thus provides "objective reassurance that the joke-work has been successful" and realizes the joker's own libidinal energy through feedback. The subject-aeffect caused by joking always relies on a referent-we to which it can be addressed. Freud's third person, the (imagined) listener/spectator, "bribed by the effortless satisfaction of his own libido," does not need make any effort whatsoever (*Joke* 95). The tendentiously propertizing speech act of the first person is directed at the third (at the cost of the second, propertized one) to corrupt them with the currency of psychical expenditure saved without effort. Given a shared libidinal ground, the third person will automatically ally themselves with the first person against the second person, who is performatively (re-)created as a *communally shared obstacle*, becoming the abjected non-reference for the communally celebrated and indulged subject-aeffect. On the ground of the second person, the community is, in fact, (re-)created.

This is also the sociogenic function of humor on the improv stage. The individual white subject makes a joke or performs an act of anti-Black abjection for humorous pleasure, thus bribing the audience – should there be a need for such

6.2 Humor and the libidinal economy of anti-Black abjection — 217

a bribe – to join in the fun because the effort of performance has already been made *for* them. This can be activated on any level, whether through content – as in the use of the n-word or defining the scene partner as a Slave – or through subliminal effects like excluding Black-racialized improvisers from protagonist roles in collectively improvised plays (treatment *like a* prop), or in micro-aggressive, non-linguistic expressions of white subject-bodies perceptible to both audience and Black-racialized scene partners. These dimensions are not necessarily voluntary but are often the result of a libidinal structure that white subjects cannot easily and intentionally *will away*, especially not in the heated situation of live public improvisation. This is especially true in improv because improvisers perform as themselves, from their own knowledges. They speak to and laugh with their audience, intuitively tailoring their language and content to that audience, as in any other form of communication. To some degree, the improviser (unconsciously) knows or assumes a certain way of being in the world on the part of their audience, and then plays to that shared knowledge – intuitively and therefore abjectively when it comes to anti-Blackness.

Given the powerful biochemical processes of laughter (in the release of endorphins), humor is not only analogical to but actively complicit in the sociogenic procedures by which nongenetic codes (mythoi) are "neurochemically implemented," as Wynter writes ("Catastrophe" 27). The central positions of the first and the third person, as well as their interplay, can be logically related to Wynter's elaborations on the imaginary referent-we of those speaking and spoken to with the aim of *kin-recognition:*

> This dynamic emerges, for example, in the "imagined communities" of our respective ethno-class nation-states: the genre-specific subjects of each such nation-state are enabled to subjectively experience themselves/ourselves in fictively eusocialized terms [...] as inter-altruistic kin-recognizing member subjects of the same *referent-we* and its imagined community. As such, kin-recognizing member subjects lawlikely and performatively enact themselves/ourselves as "good men and women" of their/our *kind* according to *nongenetically* determined, origin mythically chartered symbolically encoded and semantically enacted set of symbolic life/death instructions. ("Catastrophe" 27)

Leaving aside Wynter's focus on nation, the idea of an imagined community can be linked to the actual theatrical community of improv. Improvisers are speaking – though through their fictionalized words and actions – directly to the audience. Only if there is intelligibility among them will they recognize each other as *kin* by laughing at the same thing. The white improviser can be a "good man or woman" of their kind by making the effort of anti-Black abjection *for* them, providing them with effortless jouissance on the shared ground of the modern, racialized, and specifically anti-Black life/death dichotomy. The improvised abjection strengthens the (collectively imagined) referent-we, keeping it symbolically alive by the continual abjection of Blackness and those racialized as Black, who are discursively designed as already socially and symbolically dead. Given its material, affective, biochemical dimensions, laughter can play a powerful role in the neurochemical implementation of the social codes in biological bodies, in sociogeny. Humor and laughter program us for further humor and laughter, creating an affective and behavior-motivating autobahn that gets us *there* fast – "there" being the jouissance of overcoming an obstacle in order to relieve tension. This tension can be set up, as in a crafted joke with a punch line, or it can work as pure abjective tendentiousness. Staged moments of anti-Black abjection realized through the third person's laughter is thus pure sociogeny. Even if an improvising subject does not know if something was funny, the audience (even if only imagined) will let us know it was. The improvisers present themselves as good ones of their kind and the loudest, the most compulsive audience reactions reassure them that they are. We can be good ones of our kind only if we communicate as clearly as we can the already existent libidinal ground of the very culture or community to which we imagine we belong.

As suggested above, by way of conceptual analogy and performative likeness, the theatrical performance of the joke can be read through the communicational mechanisms of property claimantship. Rose makes the point that "in defining the acts of possession that make up a claim to property, the law not only rewards the author of the 'text'; it also puts an imprimatur on a particular symbolic system and on the audience that uses this system" (85). Like a property claim, the abjective, tendentious joke functions as a text performed in a social context. Blackness, whether in the form of an actual Black-racialized improviser on the stage or in fictive content that involves Blackness, serves white subjects by powering the anti-Black discursive configuration in which these speech acts can be uttered, and thus the culturally specific modern sociability to which they give rise. The linkage between white sociality as anti-Black abjection qua propertization and laughter can be made even more concrete in considering another key text in humor theory: Henri Bergson's *On Laughter*. In the following section, I consider the theoretical, historical groundwork of Human humor, demonstrating how the notion of *humorously dehumanized Blackness* has been mobilized within the Humanist theorization of humor – and the assumptions and observations on which these theorizations have been based.

6.3 Laughable Blackness

Psychosocial harmonization

It may seem strange to consider Bergson's largely unfashionable superiority theory of humor, which finds probably its sole contemporary proponent in Roger Scruton.[3] Concepts of superiority do not go down well in the age of mainstream multiculturalism, relativist ideology, and fetishized difference. It is not particularly fashionable (at least not in the realm of left liberalism) to take seriously concepts of superiority in popular culture or aesthetic analysis. Derogatory laughter is not something to which contemporary theorists readily subscribe, nor are they happy to accept symbolic hierarchization among humans, even though it axiomatically underlies most of the concepts applied to discuss humor. I revisit Bergson here for two reasons. First, he is one of the most influential humor theorists and one of the earliest to devote a long piece of writing to the subject, rather than using humor to illustrate other ideas (as did Plato, Hobbes, or the Christian thinkers). His essay *Le Rire* (*Laughter*) was originally published in 1900, five years before Freud's writings on jokes.[4] Bergson is thus one of the founding figures of contemporary humor studies and is still continually referenced in the scholarship. Second, his essay on humor stands within his influential vitalism at large, thus drawing a connection between these two investigative frames. For Bergson, laughter is necessarily collective:

> You would hardly appreciate the comic if you felt yourself isolated from others. Laughter appears to stand in need of an echo, Listen to it carefully: it is not an articulate, clear, well-defined sound; it is something which would fain be prolonged by reverberating from one to another [...] Our laughter is always the laughter of a group. [...] However spontaneous it seems, laughter always implies a kind of secret freemasonry, or even complicity, with other laughers, real or imaginary. (*Laughter* 11)

From this, we can extrapolate three elements: a) Laughter is an endeavor of agency-endowed cognizers, i.e., *human agents*, engaging in b) a *communicative process*, which creates c) a *community* of laughing agents collectively reacting to an object perceived as humorous. This is completely congruent with the performance situation as well as with the communicative dimension of the Freudian humorous setup involving the third person, the communicational function of

---

[3] The highly conservative and often controversial late philosopher develops the notion of "attentive demolition" in his humor theory, which resonates with Bergson and Freud (Scruton 169).
[4] Freud praises *Le Rire* as "attractive and lively" (*Joke* 214).

the prop according to Sofer,[5] the communicational ground for (anti-Black) property claims, and Wynter's concept of a (potentially imagined) referent-we.

Bergson's explicit inclusion of a potentially imaginary collectivity is worth noting in that it allows humor to take place purely in the mind of an individual. An actual group of people need not be present; it suffices that the individual understands themselves as a member of a particular cultural group, assuming that others share the same code and libidinal structure. Laughter is thus an effect of the continual, societal, discursive, individually and culturally solipsistic auto-re-institution of a community communicating with itself. Accordingly, Bergson's primary interest lies in the actual situation of humorous performance of sociality:

> To understand laughter, we must put it back into its natural environment, which is society, and above all, we must determine the utility of its function, which is a social one [...] Laughter must always answer to certain requirements of life in common. It must have a *social* signification. (*Laughter* 12)

Like the Freudian joke, Bergsonian laughter in a theatrical setup presents a situation that is both aesthetic and real, in which fictions and fantasies work with and against the actual and real that can be seen on stage. Because laughter in the theatrical situation is such a condensed and public display of what is (believed to be) collectively understood as funny, it is a meaningful site to learn about the significatory power and exertion of anti-Black force. Bergson writes further:

> Laughter [...] does not belong to the province of esthetics alone, since unconsciously (and even immorally in many particular instances) it pursues the utilitarian aim of general improvement. And yet there is something aesthetic about it, since the comic comes into being just when society and the individual, freed from the worry of self-preservation, begin to regard themselves as works of art. (*Laughter* 17)

If we think about the aesthetic and material dimensions of play, improvisation, and theater, they emerge as primary sites for humor. The reverse is also true; wherever a situation turns humorous, we see Humans at play, performing in a

---

[5] One of the central functions of the prop, as Sofer argues, is that it establishes a temporal contract. The prop embodies "a volatile 'temporal contract' between actor and spectator for the duration of performance" (ix). This means that what a specific entity signifies on a stage is decided between the improviser who offers up that signification and the audience that accepts or rejects it, or a standardized agreement that can safely be assumed to be shared right away. It is thus a central moment in actor-audience communication, providing its medium and matter.

theatrical situation of some sort whatever the context or analytic frame. Whether with Bergson's sociality, Freud's third person, or Wynter's culturally specific referent-we, humor and laughter are always located in the sphere of theatricality, including its sociopolitical bearings and libidinal structures.

Racialized rigidity

Bergson reads a corrective function in laughter, which is – sociogenically – supposed to optimize both the individual and the collective:

> In a word, if a circle be drawn round those actions and dispositions – implied in the individual or social life – to which their natural consequences bring their own penalties, there remains outside this sphere of emotions and struggle – and within a neutral zone in which man simply exposes himself to man's curiosity – a certain rigidity of the body, mind and character, that society would still like to get rid of in order to obtain from its members the greatest possible degree of elasticity and sociability. This rigidity is comic, laughter is corrective. (*Laughter* 17)

It should come as no surprise that this rigidity is racialized as Blackness. At the core of his theory of laughter lies the same Manichean binary that structures his vitalist philosophy writ large: the dichotomous descriptive statement that posits white, rational, spiritual, and cognitional symbolic life vs. Black, irrational, symbolic death. In his view, laughter is a cultural mode that serves a society's ultimate (natural, teleological) cause by reasserting its ideals in order to ensure its expansion and growth. As a gesture of social sanctioning, laughter helps to correct any aberration from social ideals, so he argues. Bergson describes the ideal of the individual living in an organic society as *elasticity*. He suggests, in Darwinian rhetoric, that such elasticity represents an individual subject's adaptability to social norms. He defines deviations from the ideal of elasticity as "elements of inferiority," which need to be removed in order to *live* within that society:

> What life and society require of each of us is a constantly alert attention that discerns the outlines of the present situation, together with a certain elasticity of the mind and body to enable us to adapt ourselves in consequence. *Tension* and *elasticity* are two forces, mutually complementary, which bring life into play. If these two forces are lacking in the body to any considerable extent, we have sickness and infirmity and accidents of every kind. If they are lacking in mind, we find every degree of mental deficiency, every variety of insanity. Finally, if they are lacking in character, we have cases of the gravest inadaptability to social life, which are the sources of misery and at times the causes of crime. Once these elements of inferiority are removed – and they tend to eliminate themselves in what has been called

> the struggle for life – the person can live, and that in common with other persons. (*Laughter* 16–17)⁶

Social, symbolic life is ensured by the individual's ability to adapt quickly to social demands. The ultimate Other to this constantly adaptive, elastic Human mind is the individual whose body, mind, or character is too static to "conduct in accordance with the reality which is present" (*Laughter* 13). Such absentmindedness (a state of mechanical lifelessness) in the Humanoid objects of humor is central in Bergson's theory: Human acts are perceived as comical if they are somehow automatic and mechanical, demonstrating a lack of self-awareness. Humorous situations include physical automatisms like tripping over a stone, and psychic automatisms such as "mathematical punctuality." In either case, the subject is not in control, not in possession of itself (like Adorno's jazz-subject), and acts like an anthropomorphic automaton, with no mind of its own and no self-awareness.

Even though this notion of a Human subject must be read as always already raced, let us follow Bergson's argument to the end. In his elaboration, the nonplus-ultra of mindless rigidity, the absolutized Other of the flexible and adaptive Human spirit – a transumption of the Divine Spirit – is the body that the modern matrix of Human existence signifies as Black: a transumption of the Fallen Flesh. Remarking on the work of the caricaturist, who "divines the deep-seated recalcitrance of matter," Bergson argues that this "art, which has the touch of the diabolical, raises up the demon who had been overthrown by the angel." The materiality of a person's physiognomy is as comical in Bergsonian theory as are all other bodily actions that draw attention to themselves or to the body that performs them. Bergson writes that "[the] attitudes, gestures and movements of the human body are laughable in exact proportion as that body reminds us of a mere machine" (*Laughter* 20–21). Whenever physicality (the materiality of the body) takes over, whenever the mind is absent or out of control, if "*some rigidity or other* [is] applied to the mobility of life," the result, according to Bergson, is funny. Rigidity foregrounds the materiality that surrounds the human spirit. His example is fashion, which foregrounds its materiality when it is out of date:

---

6 Bergson's elaboration sounds like a description of an ideal improviser and his rhetoric – "bringing life into play" – emphasizes this echo. Elasticity as an abstract concept is one quality of a good improviser, and flexibility of the mind and elegant use of the body are central to improv stagecraft. As a result of arguments like this, vitalist ideas, however esoterically or scientifically fashioned, always appeal to improvisers.

6.3 Laughable Blackness — 223

> Suppose [...] some eccentric individual dresses himself in the fashion of former times: our attention is immediately drawn to the clothes themselves, we absolutely distinguish them from the individual, we say that the latter *is disguising himself*, – as though every article of clothing were not a disguise! – and the laughable aspect of fashion comes out of the shadow into the light. (*Laughter* 25)

Through this example, Bergson introduces his influential idea of humor as emerging from "something mechanical encrusted upon the living," from moments in which "the living body [becomes] rigid, like a machine." He states, "it seemed to us that the living body ought to be the perfection of suppleness [...] the very flame of life, kindled within us by a higher principle and perceived through the body, as if through a glass" (*Laughter* 29). Perceiving this "gracefulness [...] in the living body" means to "disregard in it the elements of weight, of resistance, and, in a word, of matter" and thinking "only of its "vitality which we regard as derived from the very principle of intellectual and moral life." Yet if our attention is drawn to the material side of the principle with which it is animated, the body is no more in our eyes than a heavy and cumbersome vesture, a kind of irksome ballast which holds down to earth a soul eager to rise aloft [...] Any incident is comic that calls our attention to the physical in a person when it is the moral side that is concerned. (*Laughter* 29–30)

For Bergson, then, humor *by its defining function* abjects physicality as static, inflexible matter that restricts the Western subject's Human potential: "The comic will come into being, it appears, whenever a group of men [sic] concentrate their attention on one of their number, imposing silence on their emotions and calling into play nothing but their intelligence" (*Laughter* 12). In his devalorization of physicality (embodiment), Bergson locates rationality, creative intelligence, mental flexibility, grace, and the capacity to participate in sociality solely in the sphere of the mind. The mind is everything; the body is nothing. He thus works within the Cartesian mind-matter distinction and argues along the lines of the "fundamental Platonic postulate (that of an eternal, 'divinized' cosmos as contrasted with the Earth, which was not only subject to change and corruption but was fixed and unmoving at the center)," later transumed in the theocentric terms of Judeo-Christian Europe (Wynter, "Unsettling" 271–72). Compared to the eternal soul, the spirit, gracefulness, and intellectual levity, the body is weighty ballast that keeps the former from fully lifting itself up to where it rightfully belongs.

At a time when the natural sciences were advancing, Bergson's vitalism reinstalled both the theocentric and the ratiocentric matrices of thought while ensuring that Blackness-as-demonic physicality kept signifying deviance and degradation. Similarly, under the general heading of "disguise," his fashion metaphor is transumed into a racially abjective argumentational gesture that serves to exemplify the larger concept of the comically rigid:

> Why do we laugh at a head of hair which has changed from dark to blond? What is there comic about a rubicund nose? And why does one laugh at a [n-word]? The question would appear to be an embarrassing one, for it has been asked by successive psychologists such as Hecker, Kraepelin and Lipps, and all have given different replies. And yet I rather fancy the correct answer was suggested to me one day in the street by an ordinary canny, who applied the expression "unwashed" to the negro fare he was driving. Unwashed! Does not this mean that a black face, in our imagination, is one daubed over with ink or soot? If so, then a red nose can only be one which has received a coating of vermilion. And so we see that the notion of disguise has passed on something of its comic quality to instances in which there is actually no disguise, though there might be. (*Laughter* 25 – 26)

The analogical and metaphorical use of *Black-racialized skin-as-costume* aligned with a painted nose or dyed hair might appear an accident, or a negligible issue caused by the language of the time. However, it is a troubling symptom of the way in which Bergson reformulates and reintroduces a racialized hierarchy among Humans. A small rhetorical crack is in fact the punctum that provides invaluable insight into the inner workings of his concept of humor. This is not an accident but a trope that integrates smoothly into his whole argument. Blackness for Bergson serves the same affective function in the *argumentational dramaturgy of reasoning* that it fulfils in Adorno's jazz critique. It is not just there. Here the whiteness of Bergson's putatively universal (but really culturally specific) *us*-collective, his referent-we, becomes clear. A look at the applied deixis shows that his generalized account of laughter is not only *addressed at* his white European audience, but is in itself a *performative act* of reinstating their white universalized Human subjectivity. He presumes shared knowledge, experience, and judgment between himself and his readers, a harmonizing collective of laughers (which he terms a "closed circle") laughing at Black-racialized skin. The positionality of the laughing Human is established as the white European default. There is more to say about the wider context of this cab-driver's encounter with Blackness, which I will consider in more detail below.

"They get laughs for that:" making Blackness matter

Black skin is presented as skin-*like*. It is read as human*oid*. And yet it is mobilized as a specific kind of skin that foregrounds its physicality, its materiality, its bodily rigidity, its skin-*ness*. Like a painted nose or an outlandishly outdated dress, Black skin-as-metaphor serves to illustrate static matter, which restricts the Human *within* it. It thus signals a lack-of-Humanity if we take Bergson's discussion of the caricature seriously. In this context, the consequence of Bergson's logic is that Black-racialized skin becomes an outer representation of an inner deficiency. It does not symbolize Human life because, for Bergson, it has a *thingly* quality, which creates humor when set *in contrasting relation to Humanity*. For his argument to be logical, Black-racialized skin must then stand in opposition to the living Human being. Black skin-as-metaphor is analogized with the wearer of an eccentric piece of clothing who is "embarrassed by his body, looking round for some convenient cloak-room in which to deposit it" (30). The ease with which Bergson can put forward this list of analogies makes his claim that *we* laugh when "the body takes precedence of the soul" because "a person gives us the impression of being a thing" (33) a representative example of this project's thesis. Bergson's idea of humor must be read along with the way in which Black improvisers are propertized when the actuality of their bodies is foregrounded. Freud too suggests that tendentious jokes draw attention to a given physical aspect of a person. This corresponds to the psychosocial calling out of physical features on the improv stage, which equates with collectively abjective dehumanization. Consider this anecdote by Kimberly Michelle Vaughn:

> I remember I did a scene one time with a girl. It was in a LaRonde and we're on the bed. We were talking about eating marshmallows, the most stupid thing. And then she was like, "Yeah, but you're black." And I was like, "What?" I stopped everything I was doing in the scene and focused on that. "What does that mean 'I'm black?'" I remember being mad at myself for stopping the scene in that way, but I really wanted her to feel stupid. I wanted her to go home and cry for being a dumbass. For calling me out for that. Mostly improvisers get laughs for calling me out for not being "black enough." They get laughs for that. (personal conversation)

Vaughn is called out on her Blackness and is thereby propertized, abjected, immobilized, stripped of the ability to signify anything other than that Blackness. Without her consent, her body is dehumanized, *matter-ized*, and *abjectively made* to take "precedence of the soul" – for a laugh.

Even though Blackness as a *property of laughter* and an obstacle to Humanity may have gained popular momentum for and within the specific historical background of the US, many examples of Blackness are mobilized to generate a comic effect beyond this temporal and geographical context. Consider one from a 1985 German comedy movie: *Otto – Der Film*, still one of the most successful German movies. In the film, comedian Otto Waalkes has a scene in which the protagonist, played by himself, meets a Black-racialized US soldier on the street carrying a boom box on his shoulder. Otto pays him 50 D-Mark to act as a Slave so that Otto can sell him to an elderly lady two days before the "slave trade will no longer be allowed." After the transaction, Otto, in a different costume, claims that the lady is not the rightful owner of this Slave because she cannot present the correct documents to "own a Slave." While the elderly lady's ignorance and German bureaucracy are also themes here, it would be nonsensical (that is, libidinally incomprehensible) if anti-Black abjection were not mobilized. Even if Freud states that we do not always know what makes us laugh, in this case the tendentious quality of anti-Black abjection makes the joke work. The scene in which Otto meets the Black figure, portrayed by Günter Kaufmann, is an effective example of how Blackness does not need the obvious reference to enslavement to be *funny-for*. Slavery is not the joke here but Blackness-as-such. The GI does not immediately reply to Otto's approach, and Otto thinks he doesn't understand, explaining in a mock African accent: "Black head, black belly, black feet." Looking at the still confused eyes of the nameless character, Otto takes off one of his socks and presents his own coal-colored foot, which is then met with great appreciation and understanding on the part of the soldier. Even though minstrelsy would not be on the mind of the mainstream German audience, the dehumanizing notion of *Black skin as a dead crust costume* is transatlantically intelligible as "humorous" and the idea of Blackness is demonstrably funny as such. When asked by his new owner how, "being a slave," he would like his coffee, "Herr Bimbo" provides the punchline: "black."[7] This scene is only libidinally intelligible because Blackness cannot represent Human life, just as Vaughn's experience is only possible because her Blackness bears an inherent comic potential for her scene partner and audience. If Blackness could represent Human life, Bergson's argument would not operate. The winner of the 1927 Nobel Prize not only works within the logic of white social life and Black social death; he also performs and reenacts these dialectics in his widely-received essay on humor. His theorizing is anti-Black abjection in action, and thus offers insight into how subjects of the anti-Black modern West are funny.

Fantasies

Freud summarizes Bergsonian humor theory as follows: "The cause of laughter in these cases would be the divergence of the living from the lifeless" (*Joke* 201).

---

[7] The German term "bimbo" is a false cognate in that it is not the same as the derogatory English slang term for women deemed unintelligent. The German term activates the libidinal signification of both the n-word or other derogatory terms for Black-racialized people, such as "darky." It connotes "being enslaved" as well as "natural servility," as in "I am not your Bimbo." It is used interchangeably with the n-word.

His reception of Bergson provides an effective summary of the affective, libidinal ground on which humor (theory and practice) is variously performed, and how it creates a fundamentally anti-Black, dehumanizing form of comedy culturally specific to Western modernity. Within the framework of this project, Bergson's analysis of laughter locates the pleasure of humor in the experience of the white subject as a subject in (an) order. It is the experience of performing in a theater of self-perception, self-assertion, and self-aeffection based on the idea of belonging to a community of laughers who share a cultural value system and grammar. Such sociality creates the aeffect of the subject experiencing itself as a coherent unity within this order – an order whose coherence is paradigmatically defined as and organized around acts of anti-Black abjection. Laughing with Bergson always and necessarily involves acts of anti-Black abjection. Bergsonian humor theory provides ample examples of how white Humanities are parasitic on the semantics of Black social death, as does its ongoing reception and transumption.

Bergson does grant that his approach brings about "a fresh crop of difficulties." He argues that while for "reason," statements like "A red nose is a painted nose" or "a [n-word] is a white man in disguise" are absurd, "they are gospel truths to pure imagination." I transpose what Bergson describes as *fantasy* into a collective-affective linguistic signification, framing it as a sphere of internalized, embodied, and culturally specific knowledge. He writes: "So there is a logic of the imagination which is not the logic of reason, one which at times is even opposed to the latter, – with which, however, philosophy must reckon." Significantly, Bergson understands this imagination as a collective rather than individual fantasy: "It is something like the logic of dreams, though of dreams that have not been left to the whim of individual fancy, being the dreams dreamt by the whole of society." As an ideological structure of feeling, a white libidinal economy could hardly be better exemplified. This Bergsonian imagination is not a source for true creativity but speaks to a socially coded anti-Black matrix that works through humorous affects. It is individual and social, biochemical affect as well as code, bios as well as mythoi. *Laughter and humor are sociogeny.* Addressing their culturally specific whiteness cannot be theorized without recognizing their anti-Blackness, as exemplified by Bergson's theorization. Much like Adorno, Bergson was both partaking in and onto something. Still, his theory is too racially restricted (and universalist) to allow him to comprehend its full cultural implications. For Bergson, the Black-racialized person simply is ontologically funny for the universalized, white supremacist referent-we of Western modernity: "A man in disguise is comic. A man we regard as disguised is also comic. So, by analogy, any disguise is seen as comic, not only that of a man, but of that society also, and even the disguise of nature" (*Laughter* 26). If Black-racialized skin could – discursively – mean anything other than a *dead crust costume*, Bergsonian humor theory would collapse, its logic dissolve, its seemingly unraced elements lose their hold.⁸

6.4 Incongruities

Foundational Manicheanism

Even though *Le Rire* is a landmark in the development of humor studies, it was not developed in a void. Here I briefly consider the category of humor theories labelled "incongruity," including early theorists of humor Francis Hutcheson and Herbert Spencer and psychologists Kraepelin and Lipps, to whom Bergson and Freud explicitly refer. This little detour shows how anti-Blackness has always been instructive for the generation of humor theory even under the abstractified terms of a general "incongruity." Hutcheson is one of the most prominent forefathers of what came to be called "incongruity theory." For him, incongruity is not yet an abstract concept, but is built on the binaries of modern self-making: "According to Hutcheson, the cause of laughter resides in contrasts such as between 'grandeur, dignity, sanctity and perfection and ideas of meanness, baseness, profanity'" (Carroll 17). Hutcheson, an early abolitionist, drew on the modern transumption of Heavenly perfection and Earthly imperfection. His incongruity is therefore not abstract but decidedly hierarchical. Even though Blackness does not feature overtly in Hutcheson's writing on humor, he is attached to the Platonic postulate and transumptive chain of descriptive statements that followed it. His work is taken up by Herbert Spencer, who writes in the same vein: "Laughter naturally results only when consciousness is unawares transferred from great things to small – only when there is what we may call a descending incongruity" (206). The points of contact between Spencer and Freud are obvious; Spencer was the first to describe humor as an "economical phenomenon" in the service of regulating "psychic energy." Like Freud, Spencer also conceived of psychic energy in physiological terms (206).⁹ This strand of

---

8 In anthologies of humor (such as *Texte zur Theorie der Komik* (2005), regularly used in German universities), the section in which these underlying racial axioms are laid out is edited out.
9 "Among the several sets of channels into which surplus feeling might be discharged, was named the nervous system of the viscera. The sudden overflow of an arrested mental excitement, which, as we have seen, results from a descending incongruity, must doubtless stimulate not only the muscular system, as we see it does, but also the internal organs; the heart and stomach

humor theory has since become generalized and abstractified, so that incongruity denotes general "mismatches" between objects and the concepts they represent, scripts (Raskin), or second-degree concepts. The list of incongruity theorists is long and includes Schopenhauer, Kant, and Raskin, and more recently Marteinson and Arthur Koestner, among others. Koestner's frame-oriented concept of bisociation, as developed in *The Act of Creation,* is widely discussed. He conceives of humorous collision when the humorist joins "two incompatible matrices together in paradoxical synthesis." The audience has its "expectations shattered and its reason affronted [...] instead of fusion, there is collision; and in the mental disarray which ensues, emotion, deserted by reason, is flushed out in laughter." Here Koestner specifies emotions of "the self-assertive, aggressive-defensive type," which links his thought to Freud's notion of the tendentious joke and to my reading of it as gratuitous abjection (94–95). And yet, how has this incongruity originally been grounded?

In mobilizing the trope of a white carriage driver and his Black-racialized fare, Bergson positions himself in a trajectory that includes Kraepelin and Lipps as two theorists among others, all of whom were outstripped by the carriage driver whose undifferentiated reaction to a Black-racialized customer drives his own abjective and absolutist theory. Under closer investigation, however, it can be seen that Bergson does not differ significantly from his predecessors. For Bergson, it is important that the bemused cab-driver does not differentiate between whether Black-racialized skin is *natural* or was *put on like a disguise* by a white subject. In the sphere of superiority, this misfit designates a fundamental inferiority because the white subject can read Blackness only as a dead crust on the living. In this section, I will consider those writers whom Bergson sought to distinguish himself from when he wrote that "successive psychologists such as Hecker, Kraepelin and Lipps [...] all have given different replies" to the question why one laugh[s] "at the [n-word]" (*Laughter* 41).

In the essay "Zur Psychology des Komischen" (1885), Emil Kraepelin considers an "intellectual contrast" (132) as the fundamental concept underlying humor, which combines three kinds of comedy: visual comedy, situation comedy, and wordplay/jokes. Of these, visual comedy is presented as the "most elemental form of intellectual contrasts." The central idea is that a simple external stimulus can partly be explained by the subject within the concepts and terms of earlier experience, while other parts of that stimulus are perceived in sharp contrast to them. Kraepelin conceives of the perceptive moment as pure and sensual. The

---

must come in for a share of the discharge. And thus, there seems to be a good physiological basis for the popular notion that mirth-creating excitement facilitates digestion" (Spencer 207).

humorous, then, contrasts with "our treasure of imagination [...] without further intellectual processing" (134). The judgment is therefore intuitive in the strictest sense of the term:

> So it happens that to the child's experience, everything new or unfamiliar seems very light – if not outweighed by fear – e. g., the dad in a new suit, a lady in ballroom, a doll with real curly hair, a parrot, etc. The peasant laughs at the [n-word] he sees for the first time; he laughs at the art rider and the ballerina, sights that we have long since got used to. (134, my translation)[10]

In this excerpt, Blackness again serves as a marker of absolute Otherness to Humanity – on the same level as a new suit on a man, a new dress on a woman, a dead object with real hair, and an animal that speaks. Blackness needs no additional element to be in humorous contradiction with Humanity, to be human*oid* – it signifies in itself non-Humanity in the appearance of Humanity. Kraepelin's concept of visual comedy relies on an ultimate difference within the mind of the laughing cognizer. The underlying axiom recalls Bergson: Blackness in itself is conceived of as just as funny as a rigid, Humanoid prop that cannot move on its own. Kraepelin's mobilization of the n-word assumes that the white cosmopolitan self has already gotten used to seeing Blackness, whereas the rural farmer still finds it funny at first sight. In his essay "The Racial Ruse: On Blackness and Blackface Comedy in *fin-de-siècle* Germany," Jonathan Wipplinger relativizes the application of the farmer-encounters-Blackness trope by pointing out that in Kraepelin's argument, the trope allegedly does not play solely on the Black-white-binary, but also between rural and urban space in that it locates the farmer in urbanity, where it was actually possible for him to encounter Black-racialized people:

> The farmer's laughter [...] has been made possible through the growth of urban exoticism such as *Völkerschauen* of Hagenbeck, but also of African, African American and blackface performers in the variety theater. [...] Laughing at the "[n-word]" [...] emerges within Kraepelin's text as a rite of initiation into modern urban space, a space marked by entertainment and Blackness. (463)

---

10 "So kommt es, dass der kindlichen Erfahrung alles Neue, Ungewohnte sehr leicht, wenn nicht die Furcht überwiegt, komisch erscheint, z. B. der Papa in einem neuen Anzuge, eine Dame in Balltoilette, eine Puppe mit wirklichen Locken, ein Papagei usw. Der Bauer lacht über den [n-word], den er zum ersten Male sieht; er lacht über den Kunstreiter und die Ballerina, Anblicke, an die wir uns längst gewöhnt haben."

However, rather than disproving the racism of the trope, this observation underlines the function of Blackness for modern white negotiations of subjectivity – in this case, between rural and urban spaces. It provides another example of the fungibility of Blackness to engender white-on-white negotiations around shifting formations of sociability to which Black-racialized individuals themselves have no access as subjects. By positing themselves as cosmopolitan members of the world, as opposed to those they demean as provincial and backward, white inhabitants of the "urban jungle"[11] draw symbolically on Blackness as a means to shore up their superiority, which in no way includes those racialized as Black. Notably, the notion of the "urban jungle" is still mobilized today. The Second City's *diverse* ensemble, originally termed "BrownCo," was renamed (transumed) within the logic of the topos:

> I got my good start at Second City at the comedy studies first. And then they hired me to do "Urban Twist;" they used to call it "BrownCo." It's still not a name that we like or that people appreciate. The people who do the show only name the revue. They don't name the group. So that was out of our hands. (Boyd, personal conversation)

At the time of writing, the ensemble still carries the peculiar name it was given by white theater producers.

Kraepelin does not go into the details of his own racialized axioms of difference, but psychologist and philosopher Theodor Lipps does. Lipps was one of the most respected psychologists of his time and influenced much of Freud's writing on humor and other aspects of psychoanalysis. Lipps inspired Freud's concept of laughter as a result of psychic blockage by analyzing how "consciousness becomes static and locked onto a specific object, imbuing it with cathected energy until it can be released" (Wipplinger 464). Before Freud and Bergson, Lipps announces: "The factors of psychic life are not the contents of consciousness, but the psychic processes which are in themselves unconscious" (pos. 2166, my translation).[12] Unsurprisingly, in *Komik und Humor* (1898), Lipps too builds on the farmer-encounters-Blackness trope. He compares Black-racialized skin to *non-average* physical features, like a nose that's too small or too large – the former appears to create the impression of a "lower level of intellectual life" because it looks like a child's nose, while the latter is

---

[11] Charles Mills writes: "One might argue that in the United States the growing postwar popularity of the locution of 'urban jungle' reflects a subtextual (and not very sub-) reference to the increasing nonwhiteness of the residents or the inner cities" (*Contract* 48).
[12] "Es gilt also der allgemeine Satz: Die Faktoren des psychischen Lebens sind nicht die Bewusstseinsinhalte, sondern die an sich unbewussten psychischen Vorgänge."

excrescent, inappropriate, and pointless. In either case, for our imagination, he claims that the impression of the "form" diminishes the "content," that is, "the entirety of the organism and the life that fills it" (pos. 750). By way of analogy, he brings in Blackness, combining both "type" and "skin color":

> Similarly, type and skin color of the [n-word] are laughed at by the uneducated [...] Generally speaking, the [n-word] type evokes the idea of a lower level of development; the color of the skin is, to say the least, incomprehensible to the uneducated as the color of a human body. (pos. 764, my translation)[13]

Lipps repeatedly makes the point that such racial laughter is something only the uneducated would fall for, but nowhere does he suggest that Black skin *can* signify Humanity. Rather, he builds on the notion that Black skin is an obstacle for the white person to recognize full personhood in what otherwise appears to be a Human body:

> The perception of those human body shapes that the [n-word] shares with us creates an active willingness to connect with the [n-word] body the same assumption of a physical and mental life that we cannot but ascribe to our own bodies. (pos. 989, my translation)[14]

While he repeatedly assumes this tension between possibility and impossibility can be overcome, he nonetheless uses Blackness as the argumentational ground of a generic newness, which is by definition ultimately humorous:

> [The skin of the (n-word)] is new to the child, and to the naive person in general. It has not yet become comprehensible and familiar to them as a color which, just like ours, has the right to signify humanity. And yet it does lay claim to this special dignity in the eyes of the child and the naive person. According to perception, it actually has this dignity, i.e. it has it for the perceiver at the moment he surrenders to pure perception. However, this dignity melts away as soon as the first impression is over, and with it the habit of looking at white (and white only) as human skin color takes effect. Black skin color no longer appears

---

13 "Unter denselben Gesichtspunkt stellt sich der Typus und die Hautfarbe des [n-Wort], über welchen der Ungebildete lacht. Der [n-Wort]typus erweckt allgemein gesagt die Vorstellung einer niedrigeren Stufe der Entwicklung; die Hautfarbe ist wenigstens dem Ungebildeten als Farbe des menschlichen Körpers unverständlich."
14 "Die Wahrnehmung der menschlichen Körperformen, die der [n-Wort] mit uns gemein hat, erzeugt aktive Bereitschaft, mit dem [n-Wort]körper ebendenselben Gedanken eines in und hinter den Forman waltenden körperlichen und seelischen Lebens zu verbinden, wie wir ihn mit unserem Körper zu verbinden nicht umhin kommen."

to be entitled to this claim. It appears like an external coat of paint. Comedy has come into being. (pos 1188, my translation)[15]

Skin color again signifies ultimate alterity. Wipplinger agrees that "the farmer's laughter is the result of a dialectic of worth and worthlessness, of humanity and non-humanity, of 'white' and 'black'" (464) and that for "the farmer, if not for Lipps, the question remains as to whether the Humanity of the Black man is valid or whether he is a phony, a blacked-up white man" (465). The Black-racialized man *appears Human despite his Blackness*. The moment it is *revealed* that the Black body does not *actually* house a Human mind or soul, there is the humor of relief and vice versa. The perception of a "Black Humanity" does not conform to the modern white supremacist episteme in which Blackness is bound to signify the absolute Other, providing the referential ground for the abstractified static (as Bergson would have it two years later). Indeed, the farmer's (or carriage driver's) fictional experience demonstrates precisely the crisis that humor theorists seek to solve. In so doing, they develop universalized accounts of *what is funny* and *how humor operates*. We are looking at a productive but solipsist white crisis inspired by the mere presence of Blackness. If they tackle this crisis in slightly different ways, the fundamental binary of incongruence is the same. The crisis that takes hold of the carriage driver is the same crisis in which Kraepelin, Lipps, and Bergson find themselves. The cab driver laughs. The scholars write about laughing. Both abject.

I have suggested that the influential concept of incongruity is always conceptualized and mobilized in combination with the white abjective jouissance of dehumanizing Blackness. It has, however, been continually abstractified so that the originary anti-Blackness, its affective dimension, and its relationship to superiority theories are no longer visible. In the trajectory of incongruity theory, we can observe an absolute abstraction that ends up at statements like "what is key to comic amusement is [the deviation] from some presupposed norm –

---

15 "[Die Hautfarbe des (n-Wort)] ist dem Kinde, und dem naiven Menschen überhaupt, neu, d.h. sie ist ihnen noch nicht als Farbe, die ebensowohl wie die unsrige das Recht hat, Menschenfarbe zu sein, verständlich und geläufig geworden. Darum erhebt sie doch auch in den Augen des Kindes und des naiven Menschen den Anspruch auf diese besondere Würde. Vielmehr sie hat diese Würde nach Aussage der Wahrnehmung tatsächlich, d.h. sie hat sie für den Wahrnehmenden in dem Augenblick, in dem er der Wahrnehmung hingegeben ist. Diese Würde zergeht dann aber, sobald der erste Eindruck vorüber ist, und damit die Gewohnheit, als menschliche Hautfarbe die weisse und nur die weisse Farbe zu betrachten, in Wirkung tritt. Jetzt erscheint die schwarze Hautfarbe nicht mehr als zu diesem Anspruch berechtigt. Sie erscheint wie ein äusserlicher Anstrich. Damit ist die Komik ins Dasein getreten."

that is to say, an anomaly or an incongruity relative to some framework governing the ways in which we think the world is or should be" (Carroll 17). This abstraction corresponds to other transumptions of Blackness into the invisible, paradigmatically exemplified by the German usage of the term *digger*. What about this abstractified incongruity, and what about humor theories that are so fundamentally based on it? Embodied cognition can help us integrate several things here. First, even abstract thoughts are fundamentally physical and biochemical. Claxton considers mathematics – probably the sphere where abstract objectivist truth is cherished most highly – "a world of abstract entities that make patterns" (159), but cognition remains fundamentally physical: "the superstructure of mathematics is indeed underpinned by the childhood foundation of counting one's fingers" (160). Even mathematical variables can only function on the physical ground of the body-brain's perception – meaning there is no "purely aesthetic incongruity" to provide the basis for a purely aesthetic understanding humor. In other words, there is no such thing as an innocuous joke.

"Where is your brain from?" the ambiguity of Blackness-as-superpower

I conclude this chapter by returning to the world of improv. For those regularly affected by it, it has always been obvious: the "fact of blackness" (Fanon) matters – in improv and everywhere else. In the above discussion of humor and its performative dimensions, I considered various ways to frame white reception of Black theatrical and/or comedic performance. I now take a last look at what Black-racialized improvisers see as a direct effect of their Blackness on the improv scene, both on stage and in its social dimensions. Ironically, for many, Blackness often presents itself as an involuntary superpower:

> The first time I realized that the advantage of being the one person of color was when I was doing a show in a pretty large-sized theater, and the suggestion we got was "My Way." What instantly came to my mind was Usher's biggest album, his first big R'n'B album was called "My Way," and I love Usher, so I was like "Oh, this is a scene about Usher," and everyone else was like "What?" That is what makes improvisers of color unique; they can just think like themselves and are therefore much more entertaining and way more successful. The audience was like, "Where is your brain from?" I like to play with that a lot. It's kind of an advantage. It's liberating to find that. (Bullock, personal conversation)

Traditionally progressive concepts argued in the registers of *insightful laughter, laughing at oneself, overcoming ignorance through the insight of the philosopher/comedian,* and others may be read into Bullock's sense of liberation, like the notion of the *outsider as a particularly apt comedian* who pokes fun at the majority. However, I personally cannot provide such an optimistic interpretation. In this project at large, and in the previous sections on humor and Human vs. non-Human incongruity, something else is at work. Just by *thinking like themselves*, Black-racialized improvisers prove funny to their predominantly white audience. By applying the logic of racialized incongruity elaborated by Lipps and others, one can deduce that a culturally specific white reading of Bullock's performed Usher-association (and its interpretation as humorous) is grounded in the perceived grotesqueness of a speaking Black body that appears momentarily to have Human qualities, which – on the grounds of our modern episteme of anti-Blackness – white subjects intuitively believe to know it does not have the discursive capacity to house. To articulate this experience in Koestner's theory of bisociation: white expectations are shattered, white reason is affronted, and "instead of fusion, there is collision; and in the mental disarray which ensues, emotion, deserted by reason, is flushed out in laughter" (95). In Koestner's theory, fusion in humor is impossible, and "self-assertive, aggressive-defensive" laughter only ensues from the collision of two *incompatible* matrices. Bullock assumes that the question driving the audience's reaction may be "Where is your brain from?" It might also be an aggressive and only putatively empathetic "Strange. It looks like she has a brain!" A white audience finds (racialized) humor in the incongruous moment when Black-racialized performers act as their Black selves.[16] And yet, we must not forget that Black improvisers may find a sense of liberation in recognizing and acknowledging the performative and discursive violence against them. Bullock appreciates the net result of this constellation as an advantage ("It's so easy to blow their minds. Because they don't know anything about me, anything I tell them blows their minds."). Nonetheless, an unnerving ambivalence remains in the fact that, just by talking, the Black body generates humor for its white audience.

A longer rumination by Joel Boyd about being Black in this world speaks to the entire investigative field of this project by "wallowing in the contradictions" (Wilderson, "Wallowing"):

---

16 This might also explain the tremendous current success of Dave Chappelle for a white audience. The greater the excellence and genuine admiration of his writing, the greater the incongruity becomes. Black comedic performance may have a more powerfully abjective function for its white audiences, the higher its aesthetic and performative quality is valued. This allows liberal, progressive elites to delve into anti-Blackness as well – because we can now claim that we really cherish the comedy, while in fact we cannot differentiate the comic from the comedian and are thus bound to the primary collision of matrices.

One of my old directors – he directed my first Urban Twist show at Second City – said to us: "I know a lot of you are new to the community and Second City and I just want to address any questions about what it's like to be a minority in this building." And we ended up going on this weird tangent, and he ended up saying that the worst thing that can happen to us is to be black and mediocre. You can't be black and mediocre because first all you stick out anyway because you look the way you do. And if you suck, that's obviously really bad. But if you're mediocre, if you're not the funniest person on stage, it's just hard to watch. I started to notice that, and it's true. It's almost like a handicap that you have to make a superpower. If you're the only black person in a scene, you're automatically driving that scene. You or your character has to have some opinion or emotion. For white people, it's easy to just be with the group because they don't have that other cover. They don't represent anyone else but themselves. But if you're black, you carry that history too, and you have to be conscious of that. You can't take that lightly. You can't be mediocre. You can't be just one of the group and be "good." It's fun that I already have something that sets me apart. It's not fair, but it's true. If you look different, you have to be better. It's a handicap that you need to turn into a superpower. (personal conversation)

Boyd shares this experience with Aaron Freeman, the first Black improviser on the Second City stage:

> I will tell you one skin color thing that was a *huge* deal at Second City, and one of the big, huge reasons I sucked – which is that I always, always, was representing the race [...] It wasn't just me – I was representing brown people everywhere [...] I couldn't live up to it. I couldn't carry the 40 million people on my back on stage at the Second City [...] And I was so worried about it [...] that I could never relax – I couldn't be as good as I actually am. (Freeman qtd. in Seham 28–29)

For those who believe in progress, there is little evidence of it since Freeman was hired to perform for a Second City touring company in 1976. Instead, we see 40 years of stasis within a segment of popular culture that conceives of itself as democratic, free, and progressive. For those who uphold the ideal of improv's egalitarianism, its free and fair play, this is also a game-changer. What presented itself to Boyd as a "weird tangent" speaks directly to the non-existent center of this project; these anecdotes articulate the fundamental groundswell of anti-Blackness, and need no further interpretation. "Weird tangents" will always lead to punctums that help us make sense of empirical facts we cannot even begin to understand otherwise.

Boyd also states that "If you look different, you have to be better," which can be both an unfair demand that undermines any egalitarian fictions and a productive challenge. Bullock and Perkins also address this bind:

> *Bullock:* Being an improviser of color is doing extra work. For one, you very likely will not be in a group similar to you. You're very likely to be the only person of color in your group,

in your show. It's very likely that your audience is going to be predominantly white. You are already agreeing to do extra work.

Perkins: But I think it's that because of this extra work – and I know that certain of my peers get so mad about it – but I understand why people of color progress quicker. Because they have to do more work, but it's not even spoken about.

Bullock: It's not just an improv thing. It's like in life.

Perkins: You have to be quicker. If I am always constantly paraded with these race things, I have to be able to counteract quickly. That just makes me quicker at responding. (personal conversation)

When white improvisers cherish improv for being *like life*, this is also true Black improvisers in the very different ways highlighted here. The involuntary and nonsociable position of the Black-racialized improviser, structurally forced to "turn a handicap into a superpower" in order to practice improv, is located within the abjected space of discursive, symbolic incapacity (not individual incapability). Being Black is an obstacle for *real white play* because real play is something that Humans with full subjective capacity do:

Bullock: As a black improviser, you will always have ups and downs because people will try to convince you that you're too black. No matter how good an improviser you are, even if you're on the very main stage at Second City, the top pinnacle of improv, people will be like "Erm – too black!" People will assume that you're too different, that you're pulling the race card too much, that you're exploiting the fact that you're black. It doesn't process to them. They're just like, "The only reason people laugh at you is that you make these jokes." Like you're having this bag of black people tricks. "Stop pulling out black people-tricks so you can improvise with the rest of us and not stand out!" There are people who feel that way.

Perkins: This is not a bag of tricks. This is my life, my personality.

Bullock: My blackness isn't a bag of tricks. It's a thing that happens. I cannot take it off, so I might as well embrace it all the time. I will never not be black on stage, you know. So why not play a black person? If I don't say I'm black, then maybe I'm not, but as far as I am concerned, I am black all the time. (personal conversation)

I will say one last thing about Blackness: I remain awestruck by the fact that we are living in a world where the people who inhabit the sphere of discursive incapacity come out of it so much stronger than those with all powers of subjective capacity. Throughout this project, I have offered many angles on what first presented itself as a structural absence, but at this point in the analysis, all that remains for me to do is take a deep bow at those who live Blackness despite everything – and recognize their ultimate authority on the subject.

7 In lieu of a Conclusion

7.1 Original ending

I have thought long and hard about how to finish up this project. How does one write an ending to this investigation that in no way *concludes*? Originally, I decided to end with part of an interview that has had a lasting impact on me precisely because it invokes the myriad nuances and ambivalences activated by improv's anti-Blackness. This decision was counterintuitive for me; in many ways, the content of the excerpt opposes the general argument of my project and undercuts its logic from the specific viewpoint of a Black-racialized performer. It articulates an optimism of the will that I would like to leave to the improvisers themselves. In this excerpt, Warren Phynix Johnson, who claims to enter every scene as a white character because it is (perceived as) a "neutral character," adopts a position beyond morality that can be related to Afro-pessimist axioms. However, he *simultaneously* elaborates an argument for an aesthetic transcendence of the racial matrix that enables improv's performative specificities. This move happens in the register of a politics of hope I have repeatedly dismissed. However, it is not up to me to judge the way that Black-racialized improvisers deal with and talk about racism in improv – especially not as a white man from a different continent who has spent a grand total of 13 days in Chicago. One final look, then, at a conversation that took place in a Mexican lunch bar among four improvisers – one Black, one white, one Asian American, and myself – grounded in years of professional experience, and as rich in references to Martin Luther King as Leonard Nimoy.

> Johnson: As a white person on a stage, you can do anything. You can sexually harass people. You can be racist. You can do anything as a white person.
>
> Questioner: But that's not a good thing.
>
> Johnson: It doesn't matter if it's good or not. It matters that you can do it. So what? Me being a charismatic person, period. I can get away with a lot of shit. I do get away with a lot of shit. Granted, race-wise, I get it, you can do this, you can do that. But also as a charismatic person, you can do this and do that. As an attractive person, you can do this or do that. As a person who likes Star Trek! There are so many things you can get away with. If there's a thing that you're into, and you work that thing, you can get away with it. I feel like society is so hindered by race. Fuck all race, man! It's not about race. It's about what you can do as a person. Now, if you're stuck on race – that's fine, and that's cute. But say you're coming to the guy who is stuck on race as a guy who likes Star Trek –
>
> Schleelein: I love this analogy.
>
> Johnson: Thank you very much! Granted, there is a guy who is racist, but this fucking Puerto Rican loves Leonard Nimoy. He loves DeForest Kelly like he is into this –
>
> Schleelein: James Doohan?

Johnson: Everybody likes Scotty, of course. But they are into their thing. And if you work the thing that you're into, it doesn't matter what color you are. All that matters is what you care about. The content of your character. What are you into? Sometimes when it's like "Ohhh, the black thing..." or "Ohhh minority, so-and-so-and-so-and-so." Fuck that shit! I'm here because I am into improv.

Schleelein: I have a question now. Do you think that any of the actions that you have ever done, as a quote-unquote "minority performer –"

Johnson: Get it out, you little –

Schleelein: Do you think that any of the things you have ever done on stage has changed somebody's mind about not liking black people? Do you think that you have changed a racist mi –

Johnson: Absolutely.

Schleelein: Really?

Johnson: Absolutely. Doing a show in – and I'm not saying these people are racist – in Okojobi, Iowa. I toured for about two years with Second City, and you go literally everywhere. And you go to these small towns. I had a guy tell me I was Dan Aykroyd. A white guy. Older, chunky, heavy-set gentlemen. He goes, "You're Dan Aykroyd! They cast a certain way, and you're the Dan Aykroyd guy!" You understand? It's like a style of comedy, a brand of comedy.

Arashiba: Maybe at that time you were.

Schleelein: Blackroyd. Black Dan Aykroyd. But did you change a racist mind?

Johnson: Yeah, totally! This guy is totally not going to fuck with black people. Listen, to him say to me, "You're Dan Aykroyd," made me feel good. Because this guy is a fucking racist for sure, but he's like, "I enjoy the show, and I also understand how they cast and how they work." So, if they're casting for Dan Aykroyd, if they're casting for Bill Murray, if they're casting for – who the fuck is Better Call Saul?

Schleelein: Bob Odenkirk.

Johnson: Right. If you want to put a person in a category, what matters isn't race but style of comedy, brand of comedy. So, he was just like, "Yo – you're the Dan Aykroyd of the group." Comedy transcends race. Even when people are doing horrible, racist bits, you're still transcending race because we're laughing at it.

7.2 Reality updates

I have been working on this project for eight years, with long pauses due to personal circumstances. In those eight years, the anti-Blackness of US sociability has manifested publicly in many ways: the Charleston church shooting, the election of Donald Trump as president, the Charlesville attack, the El Paso shooting, the killings of numerous unarmed Black people by police officers, and more. In response, the Black Lives Matter movement has modified how the media reports on anti-Blackness and racism, and has inspired activism of many kinds. Within parts of the white improv community, a parallel sense of unease has developed. When I approached the improvisers I had interviewed to request permission to use the material, quite a few told me that *times had changed* and they would rather not be part of the project. In contrast, in my immediate improv environment, I have been able to talk about improv's fundamental anti-Blackness without being frowned on or sneered at. Yet nothing changed at the institutional level until very recently. What might have been a communal groundswell came to the fore shortly after the murder of George Floyd on 25 May 2020. Following allegations of institutional racism by Dewayne Perkins and Aasia Bullock, among others, Second City's long-time CEO Andrew Alexander resigned in June 2020. An article on vulture.com quotes them both:

> On May 31, Second City tweeted a pro-Black Lives Matter in support of this week's ongoing protests against police brutality, a sentiment former black performers like comedian Dewayne Perkins responded to with some surprise, considering their own experiences at the theater.
> "You remember when the black actors wanted to put on a Black Lives Matter Benefit show and you said only if we gave half of the proceeds to the Chicago PD, because I will never forget. Remember when you would make black people audition for job you simply just gave to white people?," the *Brooklyn Nine-Nine* writer tweeted in a threat Thursday. "Remember when you sent a bunch of your black actors to speech therapy because you said white people didn't understand us? Remember when you told me to my face I wasn't getting hired for main stage because I wasn't 'nice' enough and kept speaking out?"
> Other performers of color described similar experiences on Twitter, including *Space Force* writer Aasia LaShay Bullock, who tweeted about the theater's alleged failure to address her by a white actor, pressuring her to perform alongside him until she was forced to quit. (Kiefer)

In a long public statement from 5 June 2020, Alexander considers the maintenance of institutional racism at Second City "one of the great failures of my life," while asserting that on stage "we have always been on the right side of the issue, and of that, I am very proud." In the letter, he also announces that the "next person to fill the Executive Producer position will be a member of the BIPOC [Black, Indigenous, and People of Color] community," a commitment that he is again "proud to make." A day after his resignation, Anthony LeBlanc was announced as interim executive until a long-term executive would be found. On 8 June, the following open letter by current and former Black Second City performers was disseminated online:

> After a meeting with several Black alumni and current Second City employees, we have come to the conclusion that the erasure, racial discrimination, manipulation, pay inequity, tokenism, monetization of Black culture, and trauma-enducing experiences of Black artists at The Second City will no longer be tolerated. We cannot and will not let this abuse continue. As the artists whose names, images, and written material you still profit off of, we demand change.

After careful review of the history of Black artists at The Second City, we call for the following:

- A thorough investigation and removal of teachers, producers, directors, and other adminstrative staff guilty of microaggressions, racial transgressions, cultural appropriation, mental and verbal abuse against the Black artists who built your stages.
- A thorough investigation organization-wide and immediate removal of anyone guilty of sexual misconduct and sexual assault.
- A revision and proper accreditation (regardless of alumni standing) of the contributions of Black artists who built your stages.

Please contact us to provide you with the names and contact information of the victims to help you in your investigation.

While we acknowledge our brother Anthony LeBlanc's new role as the first Black executive producer of The Second City, the task he has been charged with is no more than integration into a burning house.While you use him to sort out a mess decades in the making, we will also guide you in moving forward. We demand the following:

- Hiring of an outside, independent HR firm.
- Hiring of an outside BIPOC-owned Diversity & Inclusion firm.
- Hiring of a BIPOC Executive producer by a steering committee with representatives of the BIPOC LQBTIA+ community from the current student body.

As Black alumni and current employees, we feel it is our responsibility to try to keep our brothers and sisters safe. You use our names to market your business, however we cannot in good conscience recommend The Second City as an effective place for Black comedy to thrive. We understand you may need time to implement these changes. We look forward to hearing from you within 72 hours.

Signed,

Aasia LaShay Bullock	Colette Gregory	Pip Lilly
Ali Barthwell	David Pompeii	Rashawn Nadine Scott
Amber Ruffin	Dewayne Perkins	Sam Richardson
Ashley Nicole Black	Diona Reasonover	Shantira Jackson
Chris Redd	Dwayne Colbert	Tawny Newsome
Christina Anthony	Edgar Blackmon	Tyler Davis
	Lisa Beasley	(Bullock et. al)

Similar letters directed at Second City were made public by Latinx, LGBTQIA, and APIMEDA (Asian, Pacific Islander, Middle Eastern, and Desi American) improvisers. This exchange, as well as the steps planned and taken to "make Second City a purposefully anti-racist institution" (from Second City's reply to the letters), are documented on the Second City website under the 11 June 2020 "Updates from the Second City." This small archive is worth a read. In August 2020, the *New York Times* published an article by Melena Ryzik and Jack Malooley titled "Second City Is Trying Not to Be Racist. Will It Work This Time?" The article is a significant follow-up on this project and provides an extensive and detailed report on this dynamic as it plays out.

In the same week the open letter to Second City was published, Olivia Jackson started a petition against iO on change.org. It too contained several demands, among them the decentralization of "theater decision making," "more power to BIPOC," and a public acknowledgment and apology from owner Charna Halpern for the "institutional racism perpetuated at iO as well as her individual history of racism." Other demands included the hiring of an "outside BIPOC Diversity & Inclusion Coordinator," and the commitment to "a fully revised and decolonized curriculum in order to create a learning environment where Black students can thrive." Halpern responded immediately in a letter posted on *Change.org*. Echoing Alexander's letter several days prior, she apologized in emotional and personal terms:

> My heart is being pulled and broken in so many different directions right now. I have been outraged at the police brutality and the violence against the Black Lives Matter movement. My heart breaks again to see and hear the experiences of BIPOC performers that have been uncomfortable, discriminated against, pained, and felt unheard at iO. As the owner of iO I must take responsibility for the failings in every department, and for my own failings. I am sorry.
>
> I started iO 40 years ago to legitimize improv as an art form and to create a safe space for all artists to be creative. I realize now that despite my goal to foster an environment of support and positive embrace, I have not been engaged or active enough in supporting the BIPOC and LGBTQIA+ members of our community. The world has changed greatly in my time and only I am responsible for my lack of adapting with these changes.
>
> I am sorry for ever patting myself on the back for incremental change. I am sorry for ever thinking small reforms were enough to fix systemic and institutional problems in our culture. (Response)

Combined with Alexander's letter, Halpern's reply provides enough material to analyze the novel genre of forced white institutional apology blended with the emotional reassurance of one's good intentions, reminding everyone of the great work they have done in the past on behalf of *the community in general*. Despite her lengthy apology, the iO will not go down the same road of *betterment* taken by Second City. As Halpern hinted in her response letter ("The future of iO is fragile. Our forced closure caused by Covid-19 has taken a large financial toll on the business"), the theater has fallen victim to the loss of revenue caused by COVID-19. Halpern states she was unable to pay property taxes during the shutdown and is closing iO permanently. Quoted in an article from the *Chicago Tribune*, she maintains that "If it were not for the pandemic I would not be closing. I would be meeting with the protesters" (George).

What does all of this mean for the present study? Is it extremely timely or entirely out of date? In some ways, both are true. While what I have written up as a "sketchy report of Chicago improv" may have been eclipsed by reality, the broader lines of analysis are less likely to be overtaken by day-to-day political developments of and in institutionalized improv comedy. Ryzik and Malloley's *New York Times* article publicizes numerous examples of anti-Blackness in improv, and since white denial no longer holds up the way it did when I began the project, such accounts speak to the urgency of a deeper analysis of improv's persistent anti-Black structures. This is not to suggest that recent developments can fundamentally disband the discursive episteme and libidinal economy of white subjectivity that shapes improv poetics, practice, and institutionality at its core. Even though the response letter from the current management of Second City demonstrates a willingness to "tear it all down" ("Updates"), I am unsure what that might mean or entail. This does not concern the individuals or their intentions but the discursive capacity for the fundamental change necessary to reframe improv and its axioms in order to erode the very Humanity on which improv relies as the praxis of intuitive humorous play. Improv would need to become something fundamentally different. In fact, the way in which improv crystallizes profoundly Humanist axioms and practices makes it a rich sphere for future analysis.

What is so special about this performative mode that it deserves particular scrutiny in this way? Aside from the fact that it was a short personal step for me as a scholar in the Humanities and as a professional improviser, I have given the most obvious answers above: in its reinvigoration of Humanist vitalism through the elements of intuition, play, and humor, and given its position within popular culture, improv presents itself as a discursive nexus for the anti-Black modern popular. Fetishizing notions of freedom, equality, and the transcendence of the rational self, improv draws on a traditional Humanism that was going out of fashion by the mid-twentieth century. This vitalist reinvigoration of the modern Human subject makes improv an apt field for investigating this specific work. Moreover, drawing on the mode of improvisation traditionally associated with Black cultural production, improv's whiteness creates a visible disparity up for discussion: "Why is improv so white, when improvisation is known discursively as Black cultural expression?" is a question everyone can understand. However, as it turns out, improv in this project serves at times as a vehicle to discuss modern anti-Blackness and anti-Black abjection as a white self-making principle irrespective of the stage craft itself. Improv becomes a sphere of magnified white modern self-making.

7.3 Retrospective reflections on scholarship in whiteface

+++TRIGGER WARNING+++ In this final section, I reflect on the performativity of my scholarship. In this study I set out to maintain a constant alertness and critical self-reflection of the ways I am implicated in the issues I address. At the same time, I thought it best not be ever-present in the writing, so as to develop the argument itself and avoid self-centered navel-gazing. While in some sections I might have found a good balance, in others I may have tended to one side or the other. So in the final section of this study, I consider this project's performativity. Even though I seek to keep this reflection on the structural plane, readers who despise such white self-reflections may be irritated. +++TRIGGER WARNING+++

What does this project contribute? What epistemic gain does it offer? It provides a counterpoise to the recently industrious but still largely celebratory academic production on all things improv. Methodologically, it offers a meticulous yet flexible way of engaging with theory to meet the specific demands of a chosen object of study, and it exemplifies the potential of exploring the interdisciplinary space between cultural studies, critical race theory, psychoanalysis, and neuroscience. More specifically, reading Adorno's critique of jazz and US popular culture through an Afro-pessimist framework improves our understanding of the debate about his theories and positions. Yet it feels inaccurate to read this as *my contribution* because the principal labor in this regard was done by scholars like Saidiya Hartman and Frank Wilderson. More broadly, when it comes to the real issues discussed in this dissertation, attempting to pin down my contribution from the position of white authorship is pointing at a mirage. In this performative constellation of author, content, and audiences, the notion of contribution is superficial, toothless, and even complicit in veiling the fact that I as author-subject remain implicated in the problems I discuss. To me, the term "contribution" implies a sense of ownership, which falls in line with Locke's labor theory of property; he posits that if one labors on something given, that is, on apparently "unowned" resources, one can claim ownership of the product. To my mind, this idea of ownership links the *who* with the *what* the academic world knows as "contribution." But the knowledges I have mobilized and the anecdotes of my interview partners are not unowned resources, and my scholarly labor in analyzing them does not make them mine. And even if that were not the case, in view of the Afro-pessimist framework that has guided me, I couldn't even fantasize what there is to contribute *to*, because the concept of contribution also involves the chimera of development, advancement, and some sort of telos.

In the introduction, I consider Sara Ahmed's "Declarations of Whiteness: The Non-performativity of Antiracism" and Christina Sharpe's statement that

7.3 Retrospective reflections on scholarship in whiteface — 245

"the only people who can *be* Afropessimists are non-Black people" (par. 81). How does this project relate to Sharpe's declaration? What is its performative effect? What does it do? On a superficial level, my analysis erodes the violently Humanist axioms and libidinally driven assumptions that have structured improv practice and its theoretical ground. It does so by scrutinizing central tenets like vitalism, egalitarian play, romanticized intuition, and universalized humor, laying bare the ways in which anti-Black violence is an integral part of improv. In eroding knowledge, challenging naturalized axioms, questioning habitualized lines of reasoning, and contesting Humanist assumptions I draw on a variety of disciplines, authors, and approaches, most centrally on psychoanalytic concepts understood and framed as embodied. To do this work at all, I rely on concepts that radically challenge this modern Humanism. I found those in Afro-pessimism, which *facilitates* the white autocritique attempted in this project. White autocritique implies the unstated assumption that whiteness can, in fact, be eroded by white people. And yet this optimism of the will is already a luxury that comes with the property of whiteness, to use Harris's term. I use Afro-pessimism to *serve* my structurally solipsist white-on-white critique. It is weaponized to attack the Humanist episteme and what emerges from it. My project, then, inhabits the problematic it discusses and performs the same solipsist procedure it critiques. It is thus important to note again that I could not have completed this project without the Black scholarship I have cited, and the vast range Black knowledges that I have not cited. There has been no mention of Jared Sexton, Patrice Douglass, David Marriot, Selamawit D. Terrefe, Fred Moten, or the entire body of work done in Black Studies before them. These lacunae in the bibliography are not incidental. I made a conscious decision to work with white scholars like Kristeva, Broeck, Winnicott, and Freud to formulate my own attacks on whiteness and white scholarship, and on improv's whiteness in particular. In retrospect, this decision was founded on two things: a) the unease that came with working primarily with Black scholarship, which I felt to be obliterative, b) the idea that white people should somehow clean up their own backyard first before asking Black scholars for help. I must concede that each one of those defensive justifications is futile. Obliteration will always take place in a configuration like this and drawing on white scholarship in no way prevents it – quite the contrary. Further, the idea of cleaning up Humanism's backyard is bound to the hopeful fantasy that the erosion of white knowledges is a) possible and b) leads to something better. These are mistakes I have made. A different kind of citational practice is called-for.

However, so as not to linger on moral judgement calls, let me consider the structural repercussions of white scholar-subjects deploying Afro-pessimism to attack white Humanism. In the introduction I laid out how this project was instigated by an observable Black absence from improv. It then became clear to me that the object under scrutiny was not so much Blackness as it was whiteness. Whiteness took the stage where Black absence was previously present. This re-centering of whiteness did not *feel good*, but I saw no alternative that would not make me an ethnographer or explorer of Blackness. Thus I weaponized Afro-pessimism as a conceptual ground for the labor of attempting to mark whiteness by discussing the historical, discursive, and libidinal reasons for its structural unmarked-ness. But did present whiteness turn out to be an in any way more definable or distinct object of study than absent Blackness? What remains from it after an analysis like this one? One crucial privilege that comes with whiteness is that it does not need to position itself. The flipside is that, even though I personally speak about being *positioned as white*, philosophically speaking, whiteness as such is not a *position*. It is ubiquitous. Even though individual and collective, discursive and affective anti-Black abjection brings about white positionality, being white is not a position. This echoes Harris's notion of whiteness as property, as entitlement as such: whiteness is the property *regime*. Afro-pessimism, as much as it cannot position itself, also cannot position whiteness – at least as far as I can see. So while the attacks on white Humanist concepts, assumptions, and practices constitute the practical and visible labor of this project, the emotional agony and libidinal challenges that accompany it reveal that the white scholar faces an underlying (Sisyphean) struggle against this white ubiquity. Further, the imagined or actual anti-Humanist destruction of the world as we know it would leave me with nothing, in existentialist terms. I would no longer be a sociable subject. Would I be capable of dealing with that? Does the work of marking whiteness by describing its practices need to be understood as a transumption of the very maintenance of whiteness under attack? Is it inspired and fueled by anti-Black abjection-as-scholarship? Is it at all possible for a white scholar-subject to work toward their own annihilation? Or is this the only act the white subject does *not* have the capacity to perform?

I have often given up on this project, convinced that as a white man I must not (and cannot) write such a piece or engage with anti-Blackness at all. I pondered cop-outs, like writing everything under erasure, handing in 300 empty pages, or giving up entirely. Recognizing the self-indulgence or privilege in these outs, I understood that I could not disappear behind the problem or its discussion. At some point I realized that these *solutions* would only solve my specifically white issue, problem, and unease with Black absence from improv. They offered a way *for me* to imagine *my work* as *morally unimpeachable*, and so meant dodging, taking myself out of the equation, performing the white transcendence of an otherwise unresolvable issue. But as a white subject, a white improvisor, a white scholar, I embody everything I critique. I continue to be implicated on all levels. The affective sensation of having understood something, the intellectual satisfaction of having connected dots in a meaningful way, the sense of closed logic when A can be explained by relating B, C, and D, the feelings of relief when a project like this is *wrapped up:* these affects are libidinally grounded in the Humanist imaginations of coherence, closure, and relational connection. But the issues I have discussed are not resolvable within Humanist imaginings or language. There can be no intellectual conclusion, no emotional closure, no moral absolution that scholarship or any other kind of work can offer in this regard. Afro-pessimism does not offer an ending or suggest ways to get there. Quite the contrary, Afro-pessimism posits irreconciliation in that argumentational space. Can I do that too? Because: this is not an end. Taking and using the notion of a "non-ending" in this way becomes the final Afro-pessimist weapon available to the white scholar. Here, "wallowing in contradictions" (Wilderson), I profit most from white voluntarism. As a white subject, I have the capacity to take this idea for my own ends without repercussions. And indeed, because I cannot undo anti-Blackness or suggest ways to dismantle it, I make use of that option. In view of methodological and axiomatic consistency and the violent obliteration it entails, to claim that *this is not an end* is cogent at this point in the argument. At the point where it just does not go any further.

Works Cited

Abbagnano, Nicola. *The Human Project: The Year 2000: With an Interview by Giuseppe Grieco*, translated by Nino Langiulli and Bruno Martini. Brill, 2002.
Adorno, Theodor W. "Abschied vom Jazz." *Europäische Revue*, vol. 9, 1933, pp. 313–316.
Adorno, Theodor W. "On Jazz." Translated by Jamie Owen Daniel, *Discourse*, vol. 12, no. 1, 1989, pp. 45–69. *JSTOR*, www.jstor.org/stable/41389140. Accessed 26 Mar. 2020.
Adorno, Theodor W. "Perennial Fashion – Jazz." *Prisms*, translated by Samuel and Shierry Weber, MIT, 1983, pp. 119–132.
Adorno, Theodor W. "Über Jazz." *Gesammelte Schriften, Bd. 17: Musikalische Schriften IV. Moments musicaux. Impromptus,* Suhrkamp, 1997, pp. 74–108.
Ahmed, Sara. "Declarations of Whiteness: The Non-performativity of Antiracism." *borderlands*, vol. 3, no. 2, 2004. *The Eclectic*, rbb85.wordpress.com/2014/08/24/declarations-of-whiteness. Accessed 12 Apr. 2019.
Allen, James. *Without Sanctuary: Lynching Photography in America*. Twin Palms, 2000.
Armistead, Kathryn. "A Critical Examination of Freud's Scientific Premise that Ontogeny Recapitulates Phylogeny in *Totem and Taboo*." *Didache*, vol. 9, no. 2, Jan. 2010. *International Board of Education, Church of the Nazarene*, didache.nazarene.org/index.php/volume-9–2/793–0902–05-armistead-ontogeny-phylogeny-didache-9–2/file. Accessed 15 Aug. 2020.
Assmann, Enya. "Was bedeutet Digga? Wir erklären das Jugendwort!" *Netzwelt*, 20 Jul. 2019, www.netzwelt.de/abkuerzung/171622-bedeutet-digga.html. Accessed 4 Apr. 2020.
Baldwin, James. "Going to Meet the Man." *Going to Meet the Man. Stories*. Vintage, 1995, pp. 227–249.
Banes, Sally. *Greenwich Village 1963: Avant-Garde Performance and the Effervescent Body*. Duke UP, 1993.
Barthes, Roland. *Camera Lucida: Reflections on Photography*. Translated by Richard Howard, Hill and Wang, 1981.
Bechara, Antoine, Hanna Damasio, and Antonio R. Damasio. "Emotion, Decision Making and the Orbitofrontal Cortex." *Cerebral Cortex*, vol. 10, no. 3, Mar. 2000, pp. 295–307. *Oxford Academic*, doi: 10.1093/cercor/10.3.295. Accessed 15 Aug. 2020.
Bechtel, William, and Robert C. Richardson. "Vitalism." Routledge Encyclopedia of Philosophy, edited by Edward Craig, Routledge, 1998. *Taylor & Francis*, doi: 10.4324/9780415249126-Q109–1. Accessed 17 Sept. 2021.
Beginner. "Beginner – Ahnma feat. Gzuz & Gentleman." *YouTube*, uploaded by Beginner, 3 Jun. 2016, www.youtube.com/watch?v=C6_Uk_2rkQg.
Belgrad, Daniel. *The Culture of Spontaneity: Improvisation and the Arts in Postwar America*. Chicago UP, 1998.
Bennett, Michael I. J. "The Rebirth of Bronzeville: Contested Space and Contrasting Visions." *The New Chicago: A Social and Cultural Analysis*, Temple UP, 2006, pp. 213–220.
Bergson, Henri. *Creative Evolution*. Translated by Arthur Mitchell, America UP, 1984.
Bergson, Henri. *An Introduction to Metaphysics*. Translated by T.E. Hulme, Liberal Arts, 1955.
Bergson, Henri. *Laughter: An Essay on the Meaning of the Comic*, translated by Cloudesley Brereton and Fred Rothwell, Arc Manor, 2008.
Blassingame, John. *Slave Testimony*. Louisiana State UP, 1977.
Bloom, Harold. *The Breaking of the Vessels*. U of Chicago P, 1982.

Bohm, David. Interview with F. David Peat and John Briggs, 1987, www.fdavidpeat.com/interviews/bohm.htm. Accessed 15 Aug. 2020.

Boskin, Joseph. *Sambo: The Rise and Demise of an American Jester.* Oxford UP, 1986.

Boyd, Joel. *Joel Boyd TV,* www.joelboydtv.com. Accessed 20 Oct. 2019.

Brady, Nicholas. "Looking for Azealia's Harlem Shake, Or How We Mistake the Politics of Obliteration for Appropriation." *out of nowhere: Black Meditations At The Cutting Edge,* 7 Mar. 2013, outofnowhereblog.wordpress.com/2013/03/07/looking-for-azealias-harlem-shake-or-how-we-mistake-the-politics-of-obliteration-for-appropriation. Accessed 15 Aug. 2020.

Broeck, Sabine. *Gender and the Abjection of Blackness.* SUNY P, 2018.

Broeck, Sabine. "Hegelian Maneuvers: Analogizing Slavery," *academia.edu,* www.academia.edu/8208990/Hegelian_Maneuvers_Analogizing_Slavery. Accessed 12 Apr. 2020. Unpublished manuscript.

Broeck, Sabine. "Legacies of Enslavism and White Abjectorship." *Postcoloniality – Decoloniality – Black Critique: Joints and Fissures,* edited by Sabine Broeck and Carsten Junker, Campus Verlag, 2014, pp. 109–128.

Brown, Tamara L. and Baruti N. Kopano. *Soul Thieves: The Appropriation and Misrepresentation of African American Popular Culture.* Palgrave Macmillan, 2014.

Bullock, Aasia LaShay. *Aasia Lashay Bullock,* www.aasialashay.com. Accessed 13 Dec. 2020.

Bullock, Aasia LaShay et al. Open Letter to Second City. 8 Jun. 2020, www.secondcity.com/wp-content/uploads/2020/06/103810944_10157303999593342_1907019738025515492_o-1.jpg. Accessed 20 Sept. 2021.

Butler, Patrick. *Hidden History of Lincoln Park.* The History P, 2015.

Byrnes, Peter-john. "Why improv is neither funny nor entertaining, according to a stand-up comedian: The form's greatest sin is encouraging the mediocre." *Chicago Reader,* 20 Dec. 2017, www.chicagoreader.com/chicago/improv-worst/Content?oid=37064111. Accessed 30 Nov. 2019.

Caillois, Roger. *Man, Play and Games.* Translated by Meyer Barash, Illinois UP, 2001.

Caines, Rebecca, and Ajay Heble. "Prologue: Spontaneous Acts." *The Improvisation Studies Reader,* edited by Rebecca Caines and Ajay Heble, Routledge, 2015, pp. 1–5.

Campbell, Tai. "Tai Campbell: Why aren't there more black people in improv?" *The Stage,* 13 Apr. 2017, www.thestage.co.uk/opinion/2017/tai-campbell-why-arent-there-more-black-people-in-improv. Accessed 13 Dec. 2019.

Carrane, Jimmy. "Play to the Top of Your Intelligence." *Improv Nerd Blog,* 26 Mar. 2014, www.jimmycarrane.com/play-top-intelligence. Accessed 2 Apr. 2020.

Carroll, Noël. *Humor: A Very Short Introduction.* Oxford UP, 2014.

Carter, Bill. "'S.N.L.' to Add Black Female Performer." *The New York Times,* 12 Dec. 2013, https://www.nytimes.com/2013/12/13/business/media/snl-to-add-black-female-cast-member-in-january.html. Accessed 2 Apr. 2020.

Chinyere, Oliver. "Why I'm Quitting UCB, And Its Problem With Diversity." *Medium,* 20 Sept. 2015, www.medium.com/@ochinyere/why-i-m-quitting-ucb-and-its-problem-with-diversity-961f1195a790. Accessed 13 Mar. 2020.

Claxton, Guy. *Intelligence in the Flesh: Why Your Mind Needs Your Body Much More Than It Thinks.* Yale UP, 2015.

Close, Del, Charna Halpern, and Kim Johnson. *Truth in Comedy: The Manual for Improvisation.* Meriwether, 1994.

Coleman, Elizabeth et al. "A Broken Record: Subjecting 'Music' to Cultural Rights." *The Ethics of Cultural Appropriation,* edited by James O. Young and Conrad G. Brunk, Wiley-Blackwell, 2012, pp. 173–210.

Coleman, Janet. *The Compass: the Improvisational Theatre That Revolutionized American Comedy.* Chicago UP, 1991.

Csíkszentmihályi, Mihály. "A Theoretical Model for Enjoyment." *The Improvisation Studies Reader,* edited by Rebecca Caines and Ajay Heble, Routledge, 2015, pp. 150–162.

Currie, Bennie M. "Second City getting serious about diversity." *LA Times,* 25 Dec. 2002, www.latimes.com/archives/la-xpm-2002-dec-25-et-currie25-story.html. Accessed 13 Mar. 2020.

Damasio, Antonio R. *Descartes' Error: Emotion, Reason and the Human Brain.* Vintage, 2006.

Damasio, Antonio R., Daniel Tranel, and Hanna C. Damasio. "Somatic markers and the guidance of behaviour: theory and preliminary testing." *Frontal Lobe Function and Dysfunction,* edited by Harvey Levin et al., Oxford UP, 1991, pp. 217–229.

Dewey, John. *Art as Experience.* Southern Illinois UP, 1987.

Dines, Gail and Jean M. (McMahon) Humez. *Gender, Race, and Class in Media: A Text-Reader.* Sage, 2014.

Eisenberger, Naomi I., Matthew D. Lieberman, and Kipling D. Williams. "Does Rejection Hurt? An fMRI Study of Social Exclusion." *Science,* vol. 302, no. 5643, 10 Oct. 2003, pp. 290–292. doi: 10.1126/science.1089134. Accessed 15 Aug. 2020.

Emscherblut. *Emscherblut – das Improtheater.* www.emscherblut.de. Accessed 5 Feb. 2020.

Erikson, Erik H. "The Problem of Ego Identity." *Journal of the American Psychoanalytic Association,* vol. 4, no. 1, Feb. 1956, pp. 56–121. doi: 10.1177/000306515600400104. Accessed 15 Aug. 2020.

"Everyone In Improv Troupe Balding." *The Onion,* 8 Sept. 2012, entertainment.theonion.com/everyone-in-improv-troupe-balding-1819573841. Accessed 2 Feb. 2020.

Fandos, Nicholas. "Hillary Clinton Calls America's Struggle With Racism Far From Over." *New York Times,* 20 Jun. 2015, www.nytimes.com/2015/06/21/us/politics/hillary-clinton-calls-americas-struggle-with-racism-far-from-over.html. Accessed 18 Jan. 2020.

Fanon, Frantz. *Black Skin, White Masks.* Pluto P, 1967.

Fark, Reinhard. *Die mißachtete Botschaft: Publizistische Aspekte des Jazz im soziokulturellen Wandel.* Spiess, 1971.

Fields, Karen E., and Barbara J. Fields. *Racecraft: The Soul of Inequality in American Life.* Verso, 2012.

Fotis, Mark. *Long Form Improvisation and American Comedy: The Harold.* Palgrave Macmillan, 2016.

Freud, Sigmund. "Beyond the Pleasure Principle." *Beyond the Pleasure Principle and Other Writings,* Penguin, 2003, pp. 43–102.

Freud, Sigmund. "The claims of psycho-analysis to scientific interest." *The Standard Edition of the Complete Psychological Works of Sigmund Freud,* Volume XIII (1913–1914), Hogarth, 1957, 163–190.

Freud, Sigmund. *The Joke and its Relation to the Unconscious,* translated by Joyce Crick, Penguin, 2003.

Gantz, Patrick. "1.2 – Collaboration." *Improv does best*, 1 May 2013, www.improvdoesbest.com/2013/05/01/1–2-collaboration-2/#more-96. Accessed 20 Apr. 2020.

George, Doug. "Chicago's iO Theater is shutting down permanently." *Chicago Tribune*, 18 Jun. 2020, www.chicagotribune.com/entertainment/theater/ct-ent-io-theater-to-close-permanently-0819–20200618–2apv5ycvj5aeti665tg3uxagxy-story.html. Accessed 15 Aug. 2020.

Gioia, Ted. *The Imperfect Art: Reflections on Jazz and Modern Culture*. Oxford UP, 1988.

Goddard, Cliff, and Anna Wierzbicka. "Cultural scripts: What are they and what are they good for?" *Intercultural Pragmatics*, vol. 1, no. 2, 2004. pp. 153–166. *Walter De Gruyter*, doi: 10.1515/iprg.2004.1.2.153. Accessed 15 Aug. 2020.

Griggs, Jeff. *Guru – My Days with Del Close*. Ivan R. Dee, 2005.

Halpern, Charna. *Art by Committee: A Guide to Advanced Improvisation*. Meriwether, 2006.

Halpern, Charna. Response from Charna Halpern, Owner and Founder of iO. *Change.org*, 10 Jun. 2020, www.change.org/p/io-chicago-i-will-not-perform-at-io-until-until-the-following-demands-are-met/u/26936317. Accessed 16 Sept. 2021.

Harris, Cheryl I. "Whiteness as Property." *Harvard Law Review*, vol. 106, no. 8 (Jun. 1993), pp. 1707–1791. *JSTOR*, doi: 10.2307/1341787. Accessed 15 Aug. 2020.

Hartman, Charles O. *Jazz Text*. Princeton UP, 1991.

Hartman, Saidiya. *Scenes of Subjection*. Oxford UP, 1997.

Hartman, Saidiya. *Lose Your Mother*. Farrar, Straus and Giroux, 2008.

Hehman, Eric, Jessica K. Flake, and Jimmy Calanchini. "Disproportionate Use of Lethal Force in Policing Is Associated With Regional Racial Biases of Residents." *Social Psychological and Personality Science*, vol. 9, no. 4, 2018, pp. 393–401. *Sage Journals*, doi: 10.1177/1948550617711229. Accessed 15 Aug. 2020.

Hines, Will. "Improv As Religion." *Improv Nonsense*, 14 Oct. 2013, improvnonsense.tumblr.com/post/64056071405/improv-as-religion. Accessed 17 Mar. 2020.

Hoffman, Kelly A., Sophie Trawalter, Jordan R. Axt, and M. Norman Oliver. "Racial bias in pain assessment and treatment recommendations, and false beliefs about biological differences between blacks and whites." *PNAS*, vol. 113, no. 16, Apr. 2016, pp. 4296–4301. *PNAS*, doi: 10.1073/pnas.1516047113. Accessed 15 Aug. 2020.

Hook, Derek. "Racism as Abjection: A Psychoanalytic Conceptualisation for a Post-Apartheid South Africa." *South African Journal of Psychology*, vol. 34, 2004, pp. 672–703. *Sage Journals*, doi: 10.1177/008124630403400410. Accessed 15 Aug. 2020.

Hook, Derek. "'Pre-discursive' racism." *Journal of Community and Applied Social Psychology*, vol. 16, no. 3, 2006, pp. 207–232. *LSE Research Online*, eprints.lse.ac.uk/957/1/Prediscursive.pdf. Accessed 15 Aug. 2020.

Horkheimer, Max. *Briefwechsel 1913–1936*. Gesammelte Schriften in 19 Bänden, vol. 15, Fischer, 1995.

"Inside the Master Class: Black Guy Auditions." *Youtube*, uploaded by UCB Comedy, 28 Jun. 2012, youtube.com/watch?v=WMoNeHxapkY&list=PL37AF6D2B5580C8CD&index=5. Accessed 28 Jun. 2018.

The iO Theatre. "DiOversity Scholarship." *The iO Theater*, www.ioimprov.com/classes/dioversity-scholarship. Accessed 20 Apr. 2020.

"The International Society for Humor Studies." *International Society for Humor Studies*, www.humorstudies.org. Accessed 1 Apr. 2020.

Jackson, Kennell. "Introduction: Traveling While Black." *Black Cultural Traffic: Crossroads in Global Performance and Popular Culture*, edited by Harry J. Elam Jr. and Kennell Jackson, Michigan UP, 2005.

Jackson, Olivia. "I Will Not Perform at iO Until the Following Demands Are Met." *Change.org*, 8 Jun. 2020, www.change.org/p/io-chicago-i-will-not-perform-at-io-until-until-the-following-demands-are-met. Accessed 15 Aug. 2020.

Jagodowski, TJ and David Pasquesi. *Improvisation at the Speed of Life: The TJ & Dave Book*. Solo Roma, 2015.

Jagodowski, TJ and David Pasquesi. *TJ and Dave: Official Website – David Pasquesi – TJ Jagodowski*, www.tjanddave.com. Accessed 3 Jan. 2020.

Johnson, Kim H. "As Del Lay Dying." *Chicago Reader*, 3 Apr. 2008, www.chicagoreader.com/chicago/as-del-lay-dying/Content?oid=1109931. Accessed 17 Feb. 2020.

Johnston, Chris. *The Improvisation Game: Discovering the Secrets of Spontaneous Performance*. Nick Hern, 2006.

Jones, Donna V. *The Racial Discourses of Life Philosophy: Négritude, Vitalism, and Modernity*. Columbia UP, 2012.

Judy, R. A. T. "On the Question of Nigga Authenticity." *boundary 2*, vol. 21, no. 3, Autumn 1994, pp. 211–230. *JSTOR*, doi: 10.2307/303605. Accessed 26 Mar. 2020.

Jung, Carl G. "Archaic Man." *Civilization in Transition*, translated by R.F.C. Hull, Routledge, 1970, pp. 50–73.

Katz, Robert. "Second City plans training center in Bronzeville." *The Chicago Maroon*, 21 Jan. 2003, www.chicagomaroon.com/2003/01/21/second-city-plans-training-center-in-bronzeville. Accessed 19 Feb. 2020.

Kiefer, Halle. "Second City CEO Andrew Alexander Resigns Following Accusations of Institutional Racism." *Vulture*, 5 Jun. 2020, www.vulture.com/2020/06/second-city-andrew-alexander-resigns-after-claims-of-racism.html. Accessed 15 Aug. 2020.

Kim, Peter. "Comedian Talks Quitting 'Dream Job' At Second City Due To Racist Audiences." *Chicagoist*, 26 Oct. 2016, www.chicagoist.com/2016/10/26/comedian_tells_us_why_he_quit_dream.php. Accessed 13 Feb. 2020.

Koestner, Victor. *The Act of Creation*. Hutchinson, 1964.

Kopano, Baruti N. "Soul Thieves: White America and the Appropriation of Hip Hop and Black Culture." *Soul Thieves: The Appropriation and Misrepresentation of African American Popular Culture*, edited by Tamara L. Brown and Baruti N. Kopano, Palgrave Macmillan, 2014, pp. 1–15.

Kozlowski, Rob. *The Art of Chicago Improv: Short Cuts to Long-Form Improvisation*. Heinemann, 2002.

Kraepelin, Emil. "Zur Psychologie des Komischen." *Philosophische Studien*, vol. 2, 1885, pp. 128–160. *ECHO*, echo.mpiwg-berlin.mpg.de/ECHOdocuView?url=/permanent/vlp/lit3329/index.meta. Accessed 30 Mar. 2020.

Kristeva, Julia. *Powers of Horror: An Essay on Abjection*. Columbia UP, 1982.

Kross, Ethan et al. "Social rejection shares somatosensory representations with physical pain." *PNAS*, vol. 108, no. 15, Apr. 2011, pp. 6270–6275. *PNAS*, doi: 10.1073/pnas.1102693108. Accessed 15 Aug. 2020.

Kurby, Pierre. "Was bedeutet Dicker / Diggah? Bedeutung, Wortherkunft, Definition." *Bedeutung Online*, 17 Mar. 2017, www.bedeutungonline.de/dicker-diggah/. Accessed 4 Apr. 2020.
Lambert, Raymond. *All Jokes Aside: Standup Comedy Is a Phunny Business*. Bolden, 2016.
Lehrer, Jonah. *How We Decide*. Mariner, 2010.
Libera, Anne. *The Second City Almanac of Improvisation*. Northwestern UP, 2004.
Lincoln, Abraham. *Abraham Lincoln: Speeches and Writings, 1832–1858*. Library of America, 1989, pp. 74–75.
Lipps, Theodor. *Komik und Humor. Eine psychologisch-ästhetische Untersuchung*. Guth, 2017. E-book.
Lösel, Gunter. "Can Robots improvise?" *Liminalities: A Journal of Performance Studies*, vol. 14, no. 1, 2018, www.liminalities.net/14–1/robots.pdf. Accessed 15 Aug. 2020.
Lösel, Gunter. "Gunter Lösel über die akademische Beschäftigung mit dem Phänomen Impro." Interview by Claudia Hoppe, *Claudia Hoppe Impro Podcast*, no. 37, 3 Jul. 2016, claudiahoppe.com/2016/07/13/podcast-nr-37-gunter-loesel-ueber-die-akademische-beschaeftigung-mit-dem-phaenomen-impro/. Accessed 15 Aug. 2020.
Lösel, Gunter. *Das Spiel mit dem Chaos: Zur Performativität des Improvisationstheaters*. Transcript, 2013.
Lott, Eric. *Love and Theft: Blackface Minstrelsy and The American Working Class*. Oxford UP, 2013.
Lyotard, Jean-François. *Libidinal Economy*. Continuum, 2004.
Mailer, Norman. "The White Negro: Superficial Reflections on the Hipster." *Dissent*, Summer 1957, pp. 276–293, www.dissentmagazine.org/article/the-white-negro-superficial-reflections-on-the-hipster-2. Accessed 15 Aug. 2020.
Markowitz, Jonathan. *Legacies of Lynching: Racial Violence and Memory*. Minnesota UP, 2004.
Marteinson, Peter. *On the Problem of the Comic*. Legas, 2006.
McDonald, Soraya N. "'SNL' lacks diversity in the cast, and in the writers' room." *The Washington Post*, 3 Nov. 2013, www.washingtonpost.com/blogs/she-the-people/wp/2013/11/03/snl-lacks-diversity-in-the-cast-and-in-the-writers-room. Accessed 18 Aug. 2019.
McDonald, Soraya N. "Diversity problems persist for Upright Citizens Brigade comedy troupe, students say." *The Washington Post*, 22 Sept. 2015, www.washingtonpost.com/news/arts-and-entertainment/wp/2015/09/22/diversity-problems-persist-for-upright-citizens-brigade-comedy-troupe-students-say. Accessed 18 Aug. 2018.
Mills, Charles. *The Racial Contract*. Cornell UP, 1997.
Mills, Charles. "White Ignorance." *Race and the Epistemologies of Ignorance*, edited by Shannon Sullivan and Nancy Tuana, SUNY, 2007, pp. 11–38.
mirrytamalez. "My two cents about the lack of diversity in upper levels of UCB Improv classes." *Miss Adventures of Milly*, 2015, mirrytamalez.tumblr.com/post/125959024672/my-two-cents-about-the-lack-of-diversity-in-upper. Accessed 15 Feb. 2020.
Morrison, Toni. *Playing in the Dark: Whiteness and the Literary Imagination*. Vintage, 1993.
Naameh, Will. "Can Improv Make You a Better Person?" *The Skinny*, 11 Oct. 2017, www.theskinny.co.uk/comedy/opinion/comedians-on-ethics-can-improv-make-you-a-better-person. Accessed 3 Nov. 2019.
Nachmanovitch, Stephen. *Free Play: Improvisation in Life and Art: Power of Improvisation in Life and the Arts*, Tarcher Perigee, 1991.

Napier, Mick. *Improvise: Scene from the Inside Out.* Heinemann, 2004.
Napier, Mick. *Behind the Scenes: Improvising Long Form.* Meriwether, 2015.
Newberg, Andrew and Mark Robert Waldman. *Words Can Change Your Brain: 12 Conversation Strategies to Build Trust, Resolve Conflict, and Increase Intimacy.* Plume, 2013.
Noble, Ben. "Improv is my Religion." *I am making all this up*, 8 Aug. 2014, immakingallthisup.com/improv-is-my-religion. Accessed 4 Jun. 2020.
Oberle, Eric. "Jazz, the Wound: Negative Identity, Culture, and the Problem of Weak Subjectivity in Theodor Adorno's Twentieth Century." *Modern Intellectual History*, vol. 13, no. 2, Aug. 2016, pp. 357–386. *Cambridge Core*, doi: 10.1017/S1479244314000614. Accessed 15 Aug. 2020.
Okiji, Fumi. *Jazz as Critique: Adorno and Black Expression Revisited*, Stanford UP, 2018.
Orchard, Jacob and Joseph Price. "County-Level Racial Prejudice and the Black-White Gap in Infant Health Outcomes," *Social Science and Medicine*, vol. 181, 2017, pp. 191–198. *APA PsycInfo*, doi: 10.1016/j.socscimed.2017.03.036. Accessed 15 Aug. 2020.
Paetzold, Ulrich. *Kunst und Kulturindustrie bei Adorno und Habermas. Perspektiven kritischer Theorie.* Deutscher Universitäts-Verlag, 2001.
Pager, Devah, Bart Bonikowski, and Bruce Western. "Discrimination in a Low-Wage Labor Market: A Field Experiment." *American Sociological Review*, vol. 74, no. 5, 2009, pp. 777–799. *Sage Journals*, doi: 10.1177/000312240907400505. Accessed 15 Aug. 2020.
Pasquesi, David. *David Pasquesi: an improviser, voice-over guy and actor based in Chicago*, www.davidpasquesi.com. Accessed 1 Apr. 2020.
Patterson, Orlando. *Slavery and Social Death: A Comparative Study.* Harvard UP, 1985.
Perkins, Dewayne. *Dewayne Perkins: Actor, Writer, Comedian*, www.dewayneperkins.com. Accessed 13 Dec. 2019.
Prince, Jocelyn. "New theater opening to mixed reviews." *Chicago Reporter*, 1 Apr. 2003, www.chicagoreporter.com/new-theater-opening-mixed-reviews. Accessed 4 Oct. 2019.
Quijani, Aníbal. "¡Qué tal raza!" *America Latina en moviemento*, 19 Sept. 2000. *ALAI*, www.alainet.org/fr/node/104865. Accessed 15 Aug. 2020.
"r/Improv – Why Improv Is Neither Funny nor Entertaining…." *Reddit*, www.reddit.com/r/improv/comments/7lak79/why_improv_is_neither_funny_nor_entertaining. Accessed 8 Jan. 2020.
Radin, Margaret J. "Property and Personhood." *Stanford Law Review*, vol. 34, no. 5, May 1982, pp. 957–1015.
Rankin, John. *Letters on American Slavery.* Negro UP, 1970, pp. 56–57.
Rodriguez, Meredith. "Improv diversity starts at the roots." *Chicago Tribune*, 12 Dec. 2013, www.chicagotribune.com/Fentertainment/ct-ott-1213-diversity-comedy-20131212-story.html. Accessed 18 Sept. 2021.
Rose, Carol M. *Property and Persuasion: Essays On The History, Theory, and Rhetoric of Ownership.* Westview P, 1994.
Rottweiler, Hektor. (Theodor Adorno). "Über Jazz." *Zeitschrift für Kritische Sozialforschung*, vol. 5, 1936, pp. 235–259. Deutscher Taschenbuch Verlag, 1980, www.kritiknetz.de/images/stories/texte/Zeitschrift_fuer_Sozialforschung_5_1936.pdf. Accessed 24 Mar. 2020.

Rowland, Patrick. "What to Expect if you're a Black Improviser." *The Second City Network*, 24 Sept. 2013, www.secondcity.com/network/what-to-expect-if-youre-a-black-improviser. Accessed 9 Nov. 2019.

Ryzik, Melena. "Chicago Comedy Institution iO Theater Is Closing." *New York Times*, 18 Jun. 2020, www.nytimes.com/2020/06/18/arts/television/io-theater-comedy-virus.html. Accessed 15 Aug. 2020.

Ryzik, Melena and Jack Malooley. "Second City is Trying Not to be Racist. Will it Work This Time?" *New York Times*, 12 Aug. 2020, www.nytimes.com/2020/08/12/movies/second-city-black-lives-matter.html. Accessed 15 Aug. 2020.

Sawyer, Keith. *Group Creativity: Music, Theater, Collaboration*. Taylor & Francis, 2003.

Sawyer, Keith. "Group Creativity: Musical Performance and Collaboration." *The Improvisation Studies Reader*, edited by Rebecca Caines and Ajay Heble, Routledge, 2015, pp. 87–100.

Sawyer, Keith. *Improvised Dialogues: Emergence and Creativity in Conversation*. Ablex Pub Corp, 2003.

Scafidi, Susan. *Who Owns Culture?: Appropriation and Authenticity in American Law*. Rutgers UP, 2005.

Schiller, Friedrich W. *On the Aesthetic Education of Man*. Translated by Reginald Snell, Dover, 2004.

Scruton, Roger. "Laughter." *The Philosophy of Laughter and Humor*, edited by John Morreall, SUNY, 1987, pp. 156–171.

The Second City. "Diversity and Inclusion." *The Second City*, www.secondcity.com/diversity-inclusion. Accessed 13 Apr. 2020.

The Second City. "Diversity and Outreach Panel Meet & Greet." *The Second City*, 13 Oct. 2020, www.secondcity.com/shows/other/second-city-diversity-and-outreach-panel-meet-and-greet/. Accessed 15 Aug. 2020.

The Second City. "Updates from The Second City." *The Second City*, 11 Jun. 2020, www.secondcity.com/updates-from-the-second-city-2020. Accessed 16 Sept. 2021.

Seham, Amy. *Whose Improv is it anyway?: Beyond Second City*. Minnesota UP, 2001.

Semuels, Alana. "Chicago's Awful Divide." *The Atlantic*, 28 Mar. 2018, www.theatlantic.com/business/archive/2018/03/chicago-segregation-poverty/556649/. Accessed 4 Mar. 2019.

Sharpe, Christina. "What Exceeds the Hold?: An Interview with Christina Sharpe." Interview with Selamawit Terrefe, *Rhizomes: Cultural Studies in Emerging Knowledge*, iss. 29, 2016. doi: 10.20415/rhiz/029.e06. Accessed 15 Aug. 2020.

Shusterman, Richard. "Don't Believe the Hype: Animadversions on the Critique of Popular Art." *Poetics Today*, vol. 14, no. 1, 1993, pp. 101–122. doi: 10.2307/1773143. Accessed 26 Mar. 2020.

Shusterman, Richard. *Pragmatist Aesthetics: Living Beauty, Rethinking Art*. Rowman & Littlefield, 2000.

Simons, Seth. "The UCB Theatre Released and Quickly Retracted a Very Dumb Poster." *Paste Magazine*, Paste Media Group, 28 Mar. 2017, www.pastemagazine.com/comedy/ucb/the-ucb-released-and-quickly-retracted-a-very-dumb. Accessed 4 Oct. 2019.

Smallwood, Sally and Cameron Algie. "Group Mind: Leave Your Brain At The Door." *People and Chairs: The improv blog with attitude*, 7 Jan. 2014, www.peopleandchairs.com/2014/01/07/group-mind-leave-your-brain-at-the-door. Accessed 2 Feb. 2020.

Smith, Sid. "Funny Business." *Chicago Tribune*, 9 Apr. 1995, www.chicagotribune.com/news/ct-xpm-1995-04-09-9504090118-story.html. Accessed 15 Aug. 2020.

Sofer, Andrew. *The Stage Life of Props*. Michigan UP, 2003.
Spencer, Herbert. "The Physiology of Laughter." *Illustrations of Universal Progress; A Series of Discussions*. Appleton, 1867, pp. 194–209. Project Gutenberg, www.gutenberg.org/files/39977/39977-h/39977-h.htm. Accessed 1 Apr. 2020.
Spillers, Hortense J. "The Idea of Black Culture." *The New Centennial Review*, vol. 6, no. 3, Winter 2006, pp. 7–28. doi: 10.1353/ncr.2007.0022. Accessed 15 Aug. 2020.
Spitznagel, Eric. "Follow the Fear." *The Believer*, iss. 47, 1 Sept. 2007, Black Mountain Institute, believermag.com/follow-the-fear. Accessed 15 Aug. 2020.
Spolin, Viola. *Improvisation for the Theater: A Handbook of Teaching and Directing Techniques*. 1963. Northwestern UP, 1999.
Stir Friday Night! www.stirfridaynight.com. Accessed 17 Dec. 2019.
Streu, Alex. "Improv Makes You a Better Person: Improv is the best (and cheapest) therapy you could pay for." *Medium*, 16 Jun. 2019, www.medium.com/@alexstreu.coaching/improv-makes-you-a-better-person-2ed6abbfa643. Accessed 3 Nov. 2019.
Sweet, Jeffrey. *Something Wonderful Right Away*. Limelight, 2004.
Tate, Greg. *Everything But the Burden: What White People are Taking from Black Culture*. Broadway Book, 2003.
Theater Language Studio. "Why learn improvisation?" *TLS – Theatre Language Studio Frankfurt*, www.tlsfrankfurt.com/education/why-learn-improvisation. Accessed 15 Aug. 2020.
Tißberger, Martina. *Dark Continents und das UnBehagen in der weißen Kultur: Rassismus, Gender und Psychoanalyse aus einer Critical-Whiteness-Perspektive*. Unrast, 2013.
Tosches, N. *Where Dead Voices Gather*. Back Bay Books, 2002.
Tran, Simon. "Resisting Racism on the Improv Stage." *Simon Tran*, 27 Oct. 2016, www.simontran.net/blog/2016/10/27/resisting-racism-on-the-improv-stage. Accessed 5 Jan. 2019.
Upright Citizens Brigade (UCB). *Upright Citizens Brigade Training Center: Improvisation and Sketch Comedy Education in LA & NY*, ucbtrainingcenter.com. Accessed 20 Mar. 2020.
Vaughn, Kimberly Michelle. *Actor. Writer. SuburbanBlckgrl*, www.kimberlymichellevaughn.com. Accessed 16 Dec. 2019.
Velazquez, Nelson. "When Improv is the 'But' of the Joke." *Improductions, LLC*, improductionsllc.com. Accessed 3 Jan. 2020.
Waksman, Steve. *Instruments of Desire: The Electric Guitar and the Shaping of Musical Experience*. Harvard UP, 2001.
Walsh, Matt et al. *Upright Citizens Brigade Comedy Improvisation Manual*. Comedy Council of Nicea, 2013.
Warren, Calvin L. "Black Nihilism and the Politics of Hope." *CR: The New Centennial Review*, vol. 15, no. 1, Spring 2015, pp. 215–248. JSTOR, doi: 10.14321/crnewcentrevi.15.1.0215. Accessed 15 Aug. 2020.
Wasson, Sam. *Improv Nation: How We Made a Great American Art*. Houghton Mifflin Harcourt, 2017. E-book.
Watkins, Mel. *On the Real Side: A History of African American Comedy*. Lawrence Hill Books, 1999.
Wehelye, Alexander G. *Habeas Viscus: Racializing Assemblages, Biopolitics, and Black Feminist Theories of the Human*. Duke UP, 2014.

Wilderson, Frank B. *Red, White & Black: Cinema and the Structure of U.S. Antagonisms*. Combined Academic Publishing, 2010.

Wilderson, Frank B. "Wallowing in the contradictions." An interview with Percy Howard, *A Necessary Angel*, 9 Jul. 2010, percy3.wordpress.com/2010/07/09/frank-b-wilderson-"wallowing-in-the-contradictions"-part-1. Accessed 15 Aug. 2020.

Williams, Jaye A., and Frank Wilderson. "Staging (Within) Violence: A Conversation with Frank Wilderson and Jaye Austin Williams." *Rhizomes: Cultural Studies in Emerging Knowledge*, no. 29, 2016. doi: 10.20415/rhiz/029.e07. Accessed 15 Aug. 2020.

Wilson, Flannery. "How Improv Taught Me To Be A Better Person." *Medium*, 3 Feb. 2019, www.medium.com/@fwils001/how-improv-taught-me-to-be-a-better-person-d5c53d6d54e7. Accessed 2 Feb. 2020.

Winnicott, Donald W. *Playing and Reality*. Routledge, 2005.

Wipplinger, Jonathan. "The Racial Ruse: On Blackness and Blackface Comedy in *fin-de-siècle* Germany." *The German Quarterly*, vol. 84, no. 4, 2011, pp. 457–476. *JSTOR*, www.jstor.org/stable/41494686. Accessed 4 Apr. 2020.

Wittke, Carl. *Tambo and Bones*. Greenwood, 1968.

"Who We Are." *Improv for Humanity*, improvforhumanity.org/about-2/. Accessed 2 Feb. 2020.

Woods, Zach. "Zach Woods Discusses Jazz and Comedy Improv's Parallels." *Amy Poehler's Smart Girls*. Medium, 22 Jul. 2015, www.amysmartgirls.com/zach-woods-discusses-jazz-and-comedy-improvs-parallels-48a520d36dc9. Accessed 3 Jan. 2020.

Woodward, Ashley. "Libidinal Economy." *Lyotard Dictionary*, edited by Stuart Sim, Edinburgh UP, 2011, pp. 127–29.

Wynter, Leon E. *American Skin: Pop Culture, Big Business, and the End of White America*. Crown, 2002.

Wynter, Sylvia. "Ethno Or Socio Poetics." *Alcheringa: Ethnopoetics*, vol. 2, no. 2, 1976, pp. 78–94. media.sas.upenn.edu/jacket2/pdf/reissues/alcheringa/Alcheringa_New-2-2_1976.pdf. Accessed 15 Aug. 2020.

Wynter, Sylvia. "How We Mistook the Map for the Territory, and Re-Imprisoned Ourselves in Our Unbearable Wrongness of Being, of Désêtre: Black Studies Toward the Human Project." *Not Only the Master's Tools: American Studies in Theory and Practice*, edited by Lewis R. Gordon and Jane Anna Gordon, Routledge, 2016, pp. 107–70.

Wynter, Sylvia. "Toward the Sociogenic Principle." *National Identities and Socio-Political Changes in Latin America*, edited by Antonio Gómez-Moriana and Mercedes Durán-Cogan, Routledge, 2001.

Wynter, Sylvia. "Unsettling the Coloniality of Being/Power/Truth/Freedom: Towards the Human, After Man, Its Overrepresentation—An Argument." *CR: The New Centennial Review*, vol. 3, no. 3, Fall 2003, pp. 257–337. *Project MUSE*, doi: 10.1353/ncr.2004.0015. Accessed 15 Aug. 2020.

Wynter, Sylvia and Katherine McKittrick. "Unparalleled Catastrophe for Our Species? Or: To Give Humanness a Different Future: Conversations." *On Being Human as Praxis*, edited by Katherine McKittrick, Duke UP, 2015, pp. 9–89.

Ziff, Bruce and Pratima V. Rao. "Introduction to Cultural Appropriation: A Framework for Analysis." *Borrowed Power: Essays on Cultural Appropriation*, edited by Bruce Ziff and Pratima V. Rao, Rutgers UP, 1997, pp. 1–27.

Zollar, Keisha. "AIP029 | How to Navigate Diversity Issues in Improv Communities to Increase Your Acting Opportunities – with Keisha Zollar." Interview with Ben Hauck, *The Acting Income Podcast*, 14 Oct. 2015, actingincome.com/episode29. Accessed 4 Oct. 2019.

Zollar, Keisha. "There is a real problem with diversity." *Facebook*, 22 Dec. 2015, www.facebook.com/kzollar/posts/10153173848217634. Accessed 4 Oct. 2019.

Index

abjection   2, 4, 16–20, 73, 78–86, 88, 90, 92, 99f., 102–112, 114, 116–118, 122–125, 130, 132, 134, 141, 146–148, 152, 161f., 167–171, 178, 180–186, 188–190, 192, 194f., 202, 210, 212–214, 216, 218, 226f., 229, 243, 246
Adams, Samuel   164
Adorno, Theodor   18f., 111, 133, 135–149, 151f., 170
Afro-pessimism, -pessimist   4, 8–11, 16, 18, 91, 192, 238, 244–247
Ahmed, Sara   12, 244
Alexander, Andrew   36, 240, 242
Annoyance, The   24, 26, 28, 39, 42, 46
appropriation   111–115, 134
Arashiba, Dacey   28, 39, 42, 46, 49, 59, 87f., 239
Armistead, Kathryn   74
autocritique, white   2, 11, 13f., 39, 245

Baldwin, James   18, 120, 122–125
Banes, Sally   161
Barthes, Roland   5–8
Beat Generation   151, 159
Belgrad, Daniel   151f., 158–162, 166, 172, 178
Bergson, Henri   19, 22, 154–156, 170, 199, 208, 218–231, 233
bios-mythoi hybridity   16f., 71, 108
Black absence   1, 14f., 34f., 37, 39–42, 57–60, 110, 246
Bowen, Roger   34f., 128
Boyd, Joel   22, 26, 38, 49f., 52–54, 59, 183, 187, 231, 235f.
Brady, Nicholas   114f.
Broeck, Sabine   4, 16, 73, 78–80, 84, 107, 110, 119, 164
Bullock, Aasia LaShay   3, 27f., 40, 46, 48, 50f., 53, 61f., 87, 100, 197, 234–237, 240
Byrnes, Peter-john   29, 32f.

Caillois, Roger   186, 188
Chappelle, Dave   235
Chinyere, Oliver   36f., 62
Claxton, Guy   74f., 92–95, 100, 102, 105
Close, Del   20, 25, 29–34, 41, 49, 174
Csíkszentmihályi, Mihály   174f., 185
cult, improv as   29, 31–34, 40
Culture of Spontaneity   151f., 154f., 157–161, 163f., 169–172

Damasio, Antonio   17, 20, 91, 95–97, 99–106, 171, 178–180, 183
Darwin, Charles   69f., 73, 83, 221
digger   89, 163, 234

enslavement   16, 65, 70, 76, 79, 84, 87, 118, 163f., 168, 170, 198, 226

Fanon, Franz   71, 73, 76, 102, 105
flow   174f., 177
Floyd, George   240
Freeman, Aaron   236
Freud, Sigmund   21f., 71, 74f., 104, 110, 123, 202, 208–216, 219, 221, 225f., 228f., 231, 245

Ginsberg, Allen   158–160, 162
Griffin-Irons, Dionna   36, 43f.
group mind   175–177, 187

Halpern, Charna   28–30, 36, 172, 175, 242
Harris, Cheryl I.   203f.
Hartman, Saidiya   16, 18, 76, 78–80, 84, 111, 115–119, 124, 127, 162–164, 200f., 244
Hook, Derek   108
Horkheimer, Max   146
humor   4, 14, 21f., 24f., 28, 31, 34, 63f., 73, 80, 110, 123, 128, 206–209, 212f., 215–220, 222–229, 231, 233–235, 243, 245
humor, incongruity theory of   22, 228, 233
humor, release theory of   22

humor, superiority theory of   22, 208, 219
Hutcheson, Francis   228

intuition   4, 14, 19–21, 24f., 63f., 73, 80, 155f., 170–172, 174, 177, 179, 181, 183–186, 198, 206, 209f., 243, 245
iO   24, 26–29, 33f., 36, 41f., 45f., 52, 242

Jagodowski, TJ   24f., 41, 51
Johnson, Warren Phynix   28, 30, 39, 42, 49f., 87f., 238f.
jokes, innocuous   211, 234
jokes, tendentious   211–213, 215, 218, 229
Jones, Donna V.   19, 154
Judy, Ronald   19, 148

Koestner, Arthur   229, 235
Kraepelin, Emil   224, 228–231, 233
Kristeva, Julia   17, 78–85, 98f., 103, 108f., 118f., 124, 210, 213, 245

libidinal economy   24, 40, 63, 169, 171, 183
Lincoln, Abraham   117f.
Lipps, Theodor   209, 224, 228f., 231–233, 235
Lorde, Audre   9
Lösel, Gunter   23, 177
Lyotard, Jean-François   178

Markowitz, John   120f.
Mills, Charles   9, 15, 20, 47, 50, 182, 214, 231
minstrelsy   117, 119, 129–133
Morrison, Toni   33

Nichols, Mike   165, 173, 179

Oberle, Eric   136f.
Okiji, Fumi   18, 137, 140, 143, 161, 167

Pasquesi, David   24f., 41, 61
Perkins, Dewayne   3, 28, 44, 62, 87, 127, 181, 236f., 240
play   1, 4, 14, 20f., 24f., 34, 63f., 73, 80, 109f., 137, 154, 171, 173f., 179, 184–198, 200, 202f., 206, 209f., 212–214, 218, 220, 222, 236f., 243, 245

politics of hope   57–59, 62, 142, 154, 161, 183, 238
Pollock, Jackson   151, 158, 162
popular culture   111f., 115, 117, 119, 127f., 130–136, 138f., 145, 147–149
prop, definition of   199
propertization   21, 198, 201–203, 213, 218
punctum   5–8, 14, 32, 35, 37, 136, 224

racecraft   55f.
Radin, Margaret J.   21, 202, 204
Rankin, John   117f.
Rice, Thomas   129
Rose, Carol M.   21, 203
Rowland, Patrick   37–39, 44, 48f., 51, 59f., 182

Saturday Night Live   26f., 35f., 42
Sawyer, Keith   166, 174
Scafidi, Susan   112–114
Schiller, Friedrich   186
Schleelein, Derek   28, 39, 42, 46, 49f., 87, 118, 238f.
Second City   24, 26–29, 31f., 34, 36–38, 41–47, 51–53, 60f., 231, 236f., 239–243
Seham, Amy   2, 26, 30, 175–178
Sharpe, Christina   8, 244
Shusterman, Richard   18, 134f.
Sills, Paul   163f., 172
Sofer, Andrew   198–200, 203
sociogeny   64, 71f., 76, 79, 82, 91, 99, 110, 178, 214f., 218, 227
solipsism   17, 83, 98, 104, 117, 124f., 145, 164, 166, 180, 185, 193f., 196f., 208, 233, 245
somatic marker   17, 91, 101f., 104
Spencer, Herbert   228
Spinoza, Baruch de   92
Spolin, Viola   20, 161, 171–174, 186f.
studium   5–8, 14
subject-aeffect   17, 20f., 99, 102f., 107, 110, 119, 123–125, 127, 133, 147, 179, 182, 185, 188, 192, 196, 202, 213, 216

Targos, Loreen   45, 48, 54, 60f.
Tißberger, Martina   108–110

transitional blackness   110, 192, 194f., 197, 212
transitional object   20f., 110, 188–196, 210
Turner, Victor   186

Upright Citizens Brigade Theatre (UCB)   27, 32f., 36f., 52f.

Vaughn, Kimberly Michelle   32, 39, 42, 45, 47f., 53, 182, 187, 200f., 205, 213, 225f.
vitalism   19, 22, 152–157, 164, 170, 172f., 211, 221f., 243

Waalkes, Otto   225
Warren, Calvin   15, 20, 57, 142, 149
Wasson, Sam   20, 165, 168–170, 179
Watkins, Mel   117, 128–133, 139, 163, 169, 214
Wilderson, Frank   4, 15f., 54, 59, 61, 76–78, 84, 113f., 126
Williams, Jaye Austin   15, 54
Winnicott, D. W.   20f., 110, 188–198, 203, 212, 214, 245
Wynter, Sylvia   4, 16f., 20, 64–69, 71–73, 75–78, 81–83, 85, 91f., 99–101, 106, 108–110, 113, 118, 126, 129, 160, 170, 179–182, 192, 200, 210, 215, 217, 220f., 223

www.ingramcontent.com/pod-product-compliance
Lightning Source LLC
Chambersburg PA
CBHW071737150426
43191CB00010B/1611